Canadians Who Innovate

THE TRAILBLAZERS AND IDEAS THAT ARE CHANGING THE WORLD

Roseann O'Reilly Runte

Published by Simon & Schuster

New York London Toronto Sydney New Delhi

A Division of Simon & Schuster, LLC
166 King Street East, Suite 300
Toronto, Ontario M5A 1J3

This Simon & Schuster Canada edition May 2024

Simon & Schuster: Celebrating 100 Years of Publishing in 2024

For information about special discounts for bulk purchases, please contact Simon & Schuster Special Sales at 1-800-268-3216 or CustomerService@simonandschuster.ca.

Interior design by Joy O'Meara

Manufactured in the United States of America

3 5 7 9 10 8 6 4 2

Library and Archives Canada Cataloguing in Publication

Title: Canadians who innovate : the trailblazers
and ideas that are changing the world / Roseann
O'Reilly Runte.
Names: O'Reilly Runte, Roseann, author.
Description: Simon & Schuster Canada edition. | Includes bibliographical references.
Identifiers: Canadiana (print) 20230558577 | Canadiana (ebook) 20230558593 |
ISBN 9781668023853
(hardcover) | ISBN 9781668023877 (EPUB)
Subjects: LCSH: Inventors—Canada—Biography. |
LCSH: Technological innovations—Canada. | LCSH:Inventions—Canada. |
LCGFT: Biographies.
Classification: LCC T39 .O74 2024 | DDC 609.2/271—dc23

ISBN 978-1-6680-2385-3
ISBN 978-1-6680-2387-7 (ebook)

Contents

PART FIVE
Machine Learning and the Sense of Metaphor

PART SIX
The Art of Innovation and Innovating Art

PART SEVEN
Engineering Innovation

PART EIGHT
Physics, Photonics, and Nobel Laureates

Introduction

We live in a nation of innovators. Canada is a large country with a relatively small population, and when we do not have what we want, we make do or, even better, we innovate. Some of the innovators featured here are already Nobel laureates and others are on their way to fame and they just may be our next-door neighbours. Their stories are truly fascinating.

We can take pride not only in the beauty of our land and the incredible natural resources we enjoy, but also in the people who are our most important resource. These innovators are generous and humble, incredibly wise and smart. They are hardworking. Their passion for their work is infectious and inspiring. It has been an extraordinary pleasure and privilege to write about their accomplishments, aspirations, and dreams. These brilliant, successful people often faced challenges and made hard decisions—as every one of us must on occasion. They gave me great hope for the future, and it is that hope that I would like to share with you, as the path to successful innovation is open to everyone.

When it was suggested that I write a book about innovative Canadians, I said yes. After all, as president and CEO of the Canada Foundation for Innovation, this is one topic into which I might have some insight. Before beginning, however, there were two problems to solve, and both involved definitions. The first concerns nationality. I simply include people living and working in Canada or who have done so. You will find a brilliant Canadian who now lives in California and a quantum theorist who studied at the Perimeter Institute, but returned to the United States.

The second definition involves the word "innovation" itself, which is frequently linked to the application of ideas to drive business or involves the invention of technologies, processes, or products. For example, when we hear about the Canadarm on satellites, we have no trouble labelling it as an innovation. When we read about the new nasal spray vaccines that will soon be available, eliminating the need for appointments, lineups, needles, and tears, we are prompt to herald them as another innovation. However, innovation can also refer to the application of new concepts to improvements to health care or the creation of sustainable cities, to smart farms to forestry or aquaculture. It can be applied to new ways to provide housing, improve communications, or save the environment. Entrepreneurs who will create new businesses and grow the economy, researchers who will discover cures for cancer and save lives are equally innovative. Ways to solve some of the challenges the world faces today, and to create the governance models that will promote positive change in society, would all be welcome innovations.

May the individual sketches inspire you as they did me. Canada is an extraordinary country, and in this volume you will meet some of the trail-blazers whose work will change our lives and the world.

PART ONE

Innovating the Ways We Do Business

Successful business leaders frequently share the qualities that characterize innovative thinkers. They are passionate. They have invented commodities and processes. They have explored new technologies and imagined fields of endeavour. They have studied their industry and have created new businesses and new ways of doing business. They are focused, dedicated, hardworking, imaginative, and creative. They are the people who, when they are given a lemon, do not stop at making lemonade. They are the creators of lemon sorbet.

In this chapter you will discover innovative business leaders whose stories involve more than success in developing a product or a market. These trailblazers created systemic change in the way we conceive of business and its purpose.

Christina Gold cleared a path to corporate success for women to follow, and their presence in the boardrooms and their leadership in industry have made a difference not only in the world of business but also in the way we structure society and the roles to which all may aspire. Christina's innovation was inclusion, offering a new model of leadership for herself, her team, and us all.

Tom Jenkins's innovations accelerated the world's entry into the digital age. Without his work, machine learning would be an abstract concept instead of a burgeoning new field. Tom, however, did not stop with the development of innovative machines and processes, he thought ahead and promoted information sharing and what is today known as "open science," or free access to scientific information needed to improve our world.

Blake Goldring took philanthropy to a new level. He might well have made a donation to the military when they recruited him to serve as a volunteer leader. Instead, he decided to learn, visiting our military in Kandahar and their families, and listening to military leaders and their counterparts in business. He saw what was needed and brought people together to create solutions: building bridges between the military and civilians, providing support for children, and establishing a true community of shared interest. He continues to lead tirelessly with vision and empathy.

A common thread runs through the lives and work of Dave Smardon, Kayla Isabelle, and Ken Doyle. All are successful entrepreneurs with a passion. Dave's focus is on agriculture, Kayla's on supporting business that contributes to the common good, a better world, and Ken's on inclusion—especially of women and immigrants. These extraordinary leaders dedicate their lives to helping others succeed and to supporting visions of a nation where agri-tech prospers, where buildings and processes contribute to a sustainable environment, new businesses give back to those in need around the world, and everyone has the opportunity, the tools, and knowledge to succeed.

The people in this chapter are bright lights on the Canadian landscape who continue to dedicate themselves to improving our world.

1

CHRISTINA GOLD

Inspiring through Corporate Leadership

Montréal, QC

At a time when only 3 percent of the top five hundred companies were led by women, Christina Gold presided over seven major corporations, receiving many awards, including being named more than once by *Forbes* as one of the one hundred most powerful women in the world. Intelligence, the ability to include others and build teams, along with considerable hard work took this inspiring leader on the path to success.

Born in the Netherlands in 1947, Christina Gold came to Canada at the age of five, landing in Toronto, but moving to Pointe-Claire in Montréal. She could not speak English or French and found school difficult, but she developed a great interest in mathematics and geography. Due to her early language barrier, she was always afraid of not understanding others. This experience has given her a great deal of empathy for fellow immigrants. From a young age, Christina was encouraged by her parents to be self-reliant and to work hard. As a result, during school and over the summers, she worked in retail, taught swimming at a summer camp, and even read the daily pollen count on the radio.

In 1965, when it came time to consider which university to attend, Christina first settled on the city: Ottawa, which she thought was beautiful. She then narrowed her choice to Carleton, a good university with a campus she loved. She registered there as an arts student in economics and geography, and really enjoyed her studies. That first year, she also met Peter, whom she would later marry.

Upon graduation in 1969, Christina did not have a career plan, but knew that she needed to earn $125 a week to cover expenses, and "fell into a job" as a receptionist and "coupon counter." A headhunter later tried to hire her for a similar position, but she declined, saying presciently, "This job is not forever."

Shortly afterward, she saw an ad in the newspaper for an inventory accountant clerk at Avon, a company known for its door-to-door sales of cosmetics. This was Christina's big break, but she did not know it at the time. At Avon, she found a mentor, who took her under his wing, encouraging her to overcome her fear of public speaking and to seek new and challenging positions. Over the next ten to twelve years, she held nineteen different jobs in the company, eventually becoming the head of Avon Canada in 1989. In each of her positions, she successfully met every challenge through dedicated effort and built teams based on trust and respect.

In recognition of her significant achievements, Christina was then named president of Avon North America. The Canadian company was doing well, but the American company was not, and it would become Christina's responsibility to turn it around.

The new position presented some challenges. Christina had never dealt with investors from Wall Street, and when she arrived in New York, she was both an insider and an outsider. She recalls taking the elevator up to her office on the thirty-fifth floor, looking out on Central Park, and feeling somewhat terrified. But she did not let this sentiment stop her. Christina forged ahead, knowing that the key was the people: she had to respect, motivate, and make success possible for the company and its fifty thousand employees. She said her solution to making the company prosper was not only improved marketing (and her team developed a very strong publicity campaign), but a more inclusive environment that she says meant listening and communicating with employees and customers. She gave people a chance to contribute and flourish and she recognized their work, building a strong team. She diversified Avon's products, expanding into lingerie, pantyhose, and casual clothing. All the products were of excellent quality, and Christina points out that she still enjoys wearing a sweater she bought from Avon in 1992.

Her efforts bore fruit: in one year, sales increased by 10 percent and profits by 32 percent. Christina credits this successful growth to giving people opportunities. Years later, she met a woman who was a former employee who said to her, "You gave me the chance to be successful. You gave me my chance, and my product is what you sold."

Christina remarks that part of her success has also come from her travelling across North America—and around the world—learning firsthand how to work with others. When she was divisional manager of sales for Avon Canada, her district covered the area from Toronto to North Bay, Ontario. She travelled constantly, Monday to Friday, meeting regularly with several thousand door-to-door representatives. When Avon Canada became part of the Pacific portfolio, she added China and Indonesia to her travels, spending considerable time overseas. When she took the helm of Avon North America, she crisscrossed the States. Christina says that when working with employees, from fellow Canadians to colleagues overseas, it is important to respect differences and to listen and give others the opportunity to participate. This is a lesson she learned herself when trying to learn English and French to catch up with other students when she first came to Canada. As Christina says, "You cannot do it all. You are the orchestra leader, but you do not play an instrument. Your job is to help every player do their best."

In 2002, she became president and CEO of Western Union Company, a major corporation known for communications (sending telegrams) and transferring funds internationally. It operated on a network somewhat similar to that of Avon. The individual agents around the world were independent and enjoyed a culture of innovation. In 2006, with the global adoption of internet communications the company completely closed the telegram service while Western Union Financial Services remained with Christina as the president, concentrating on financial services, a direction in which it had been shifting for the last decade. Christina had been thinking of retiring, but realized there was considerable work yet to be done. She had to make everyone feel part of the whole and "build crosswalks, not silos." She brought together the hundred brightest minds in the company and instigated a thought process that worked its way through the organization.

The company had a major problem that affected total profitability, and she said that all personnel needed to kick in to solve it. So, they "held hands and jumped off the cliff together." They developed a plan with a financial strategy and details on how to communicate it around the world. The plan became everyone's strategy, all being members of the team. People were aligned with common goals.

Christina also took the time to recognize the employees who were working especially hard. Sometimes it meant simply sending a note; on other occasions, she took the time to find appropriate and thoughtful ways to appreciate others. Christina says, "It is important to say thank you." Expressing gratitude may seem like an obvious gesture now, but even into the first decade of this century, the leading business schools in America had preached that the only thing employees appreciated was their pay and that CEOs should not waste their time with anything else. For Christina to go counter to this trend and personally recognize staff members was innovative and courageous, particularly for a woman executive who could have been criticized for showing empathy.

Christina's advice is to "be curious and to seek opportunities to learn. Imagine what can be, not what cannot be. Work hard and be agile in thinking. Failure is one step on the path to success. If you take the wrong turn, pick yourself up and learn from your mistakes."

Christina believes that technology will drive much of the change in the future, but we should never underestimate the importance of people. They all matter and deserve respect, especially in this age of technology. Artificial intelligence will bring change, but will also open the door to many possibilities that companies with creative and engaged employees will discover.

Christina's leadership has been recognized by the fact that she has made significant contributions to many boards. As past chair of the board of Korn Ferry, a company that seeks to match people's talents and skills with the needs of industry, she says, "People are masters of their own destiny. We need to be enablers and help others find their career path." Christina is optimistic about the future, but cautions that we need to make sure everyone has connectivity, an "equal shot at success." Canadians must realize that we are citizens of the world and that our challenges are also global issues. As we look to solve our own problems of flooding in coastal areas, the need for clean energy, the way to produce sufficient food and to share technology with people in our North, we will gain knowledge and expertise that will provide us the opportunity to connect with people around the world and to play a role on the global stage.

Christina reflects and says that "to solve problems, young people must be curious, and we must give them a chance. We should be careful in creating legislation that will actually encourage youth participation and support the development of business." She pauses and then adds with a smile, "Fortunately, you cannot legislate away innovation." And we are fortunate indeed that Christina, one of the world's top executives, invariably leads by prioritizing people and great ideas. Christina has always opened doors, giving people the opportunity to shine, and we hope that many will be inspired to follow in her footsteps.

2

BLAKE GOLDRING

Leadership through Service

Toronto, ON

Making a difference, recognizing corporate responsibility, taking meaningful action, building bridges of hope and understanding—these phrases describe the life and actions of Blake Goldring, a strong corporate leader who has made generous giving a caring and innovative art.

Blake has served as chair of Sunnybrook Health Sciences Centre, director of the Canadian Film Centre, vice-chair of the Toronto Symphony Orchestra, and chairman of Canada Company, among a host of other directorships, including at the C. D. Howe Institute and the World Wildlife Fund. However, Blake is quick to emphasize that individuals do not need to serve on a board or offer a donation to make a difference—they just need to get involved.

Blake learned about philanthropy at home in Toronto, where he was born in 1958. His grandfather was head of the Toronto Board of Education and sat on the board of the United Community Fund. Blake's father also gave major contributions to a number of organizations and instilled in Blake the lesson that "we are blessed and fortunate to be able to give back and this is sorely needed by society."

In 1980, when he was working toward his honours degree at Victoria College, affiliated with the University of Toronto, Blake got involved in volunteer service. He was elected vice president of the Association internationale des étudiants en sciences économiques et commerciales, known by its acronym, AIESEC, an international youth-run and not-for-profit organization focused on leadership and global volunteer work. Blake ended up finishing his degree at McGill University in Montréal in order to fulfill his volunteer leadership activities with the organization, headquartered in Montréal, where it organized international exchanges for volunteers, teachers, and students to work and study with international businesses and not-for-profits. It is notable that Victoria College's motto is *Crescam Serviendo*, "Learning through Service," and the university recognized Blake's service as a volunteer lecturer and donor with an honorary degree in 2021.

On graduation, Blake went on to do his master's degree in business administration at INSEAD (Institut européen d'administration des affaires) in France, honing his language skills and continuing to form a more global perspective, before beginning his work in Toronto at the Bank of Montreal in international and corporate banking. He joined AGF Management Limited (AGF), whose name came from the American Growth Fund, the first mutual fund in Canada to invest solely in US equity in 1987. AGF itself was an asset management firm co-founded by Blake's father, C. Warren Goldring. Blake set his sights on the firm's global expansion. Under his leadership AGF was transformed, diversifying its business lines, developing new capabilities, and realizing his vision of becoming a global company while promoting community service in Canada.

Blake believes that corporations, like individual people, have a responsibility to give back to the communities in which they operate. He introduced a program at AGF called "Making a Difference." This program encouraged all employees to serve charitable organizations of their choice through both donations and paid time off to volunteer. Working with the leaders of AGF, he has helped foster a corporate environment that supports giving back to society across multiple areas, including education, the environment, and diversity, equity, and inclusion.

But perhaps Blake's greatest philanthropic success came in the early 2000s, when Canadian soldiers were sent to Afghanistan to join the international security force fighting the Taliban alongside NATO and African partners.

Blake was struck by the need to provide real assistance to these soldiers and their families, whose commitment to Canada he found so inspiring. When people volunteer for the military, they put their lives at risk to enable Canadians to enjoy life and security. Just as these soldiers stand up to protect us, Blake felt that ordinary Canadians should likewise accept some responsibility for these soldiers and their families.

A man of his word, Blake started a dialogue between military leaders and the civilian community to share lessons, and skill sets, and build bridges of understanding. It was important that civilian leaders understood the challenges faced by their military counterparts, who could, in turn, benefit from business models and practices. He quickly realized he wanted to involve more people in volunteering. Inspired by the teamwork that characterizes effective military units, Blake set out to build a team of like-minded Canadians, creating Canada Company in 2006.

Blake and the team at Canada Company established a scholarship fund for the children of soldiers who lost their lives while serving. The fund has now provided well over $774,000 in financial aid for post-secondary education. The organization also serves as a network for families of the fallen, allowing the widows and children to share their experiences and form a self-support group. It is truly moving to see the families return each year to meet at the annual scholarship awards ceremony, where the families are also welcomed to a private event when they share their personal experiences and support one another in person. They become a kind of extended family with students encouraging one another. It is truly moving to see the young scholars standing proudly on the stage as they receive their scholarships.

Since 2013, Canada Company has also helped over three thousand veterans find meaningful employment in jobs that match their skills and training at military-friendly certified companies. It sponsored camps for children of deployed parents and supported reservists and cadets through bursaries and scholarships. Additionally, the organization worked collaboratively with leaders from Canada's largest financial institutions to include a war exclusion clause in mortgage agreements to prevent foreclosure for the families of soldiers who had been sent overseas or lost their lives.

Canada Company runs numerous programs aimed at honouring the legacy of the nation's troops and celebrating local heroes across the country, as well as recognizing cadets and Junior Canadian Rangers who exemplify attributes of leadership and citizenship.

Canada Company is highly successful today, but Blake thinks back to the day he started working on it. He had been shaken by seeing firsthand the impact of the war on soldiers and their families, and this motivated him to help, but, as he humbly notes, he did not have the background or the experience to single-handedly grow an organization to the size and success that Canada Company now enjoys. At the time, he knew he had to act, but did not know if he would end up "being a one-man band pounding a drum on the street corner or if [he] would end up leading an orchestra." Today, Blake's orchestra includes volunteers and businesses from across the country. In recognition for his work, in 2011, Blake was the first person to be named Honorary Colonel for the Canadian Army.

Philanthropists work hard to find the right charitable organization to support, and do so with great care and generosity, often in tribute to a family member or for a cause about which they are passionate. What sets Blake apart is that he began by identifying a need and then created the charitable organization. He had not served in the military, but he did the research and consulted people across the country. In forming Canada Company, he created better understanding and collaboration between the military and the community. He stepped up not only as a donor but also as a volunteer to ensure the organization would carry on in the future.

Blake says that "volunteering is part of life as a good citizen. If you can give back, you have the duty to do so. We are all enriched by participating. We do not need to have a lot of money or time, but we should give what we have. In this way we learn and grow as individuals, and we further benefit from cohesiveness: sharing and growing together."

As a testament to his many contributions, Blake received the Meritorious Service Medal in 2009 and was named a member of the Order of Canada in 2018. However, he remains focused on his next steps, which he says will fall under the "rubric of education and will encourage us all to look after one another."

Blake exemplifies this heartfelt sentiment through the story of Dr. Sherif Hanna, who was a top surgeon at Sunnybrook Hospital in Toronto. In 2014, after retiring from his successful career in Canada, Dr. Hanna returned to his home country of Egypt to serve at Harpur Memorial Hospital in Menouf, a small, rural city located some two hours from Cairo. During a time of insurrection, the townspeople successfully surrounded the hospital to protect and save it from militants who wanted to destroy what was the best medical facility in the region. Sadly, Dr. Hanna passed away in 2022, shortly after returning to Canada with the intention of finally enjoying his retirement. Fortunately, the story of Dr. Hanna's efforts while he was modernizing and expanding the hospital had already led Blake to provide needed financial support not only to complete the project but also to help a fine and good man who "truly did the Lord's work, giving back, and making people healthy." One can say the same of Blake, who has quietly helped where needed, but who has also created the environment and platform from which we can all join in doing what Blake says "will make you truly happy."

3

TOM JENKINS

A Higher Degree of Innovation in a High-Tech World

Waterloo, ON

It is not surprising that as the co-author, with the Right Honourable David Johnston, of *Innovation Nation: How Canadian Innovators Made the World Smarter, Smaller, Kinder, Safer, Healthier, Wealthier, Happier*, Tom Jenkins knows a thing or two about innovation. Nor is it surprising that as the board chair and first CEO of OpenText, he is an extraordinary innovator who would have deserved a chapter in his own book.

Born in Hamilton, Ontario, in 1959, Tom says his mother strongly encouraged him to attend university. As a young lad, he took the bus to school, and every morning his mother would walk with him to his stop. On the way there, she would point at a bus stop a bit farther down the road and say, "That bus goes to McMaster University, and you are going to get on that bus one day." While Tom originally hoped to become a fighter pilot and applied to be an officer cadet in the Royal Canadian Air Force, he was unable to realize this dream because his vision was not 20/20. Thus, he did exactly as his mother suggested, and went on to earn a bachelor's degree in engineering and management at McMaster.

Tom's family could not afford to send him to university, but he was fortunate to receive a scholarship from the Natural Sciences and Engineering Research Council (NSERC). He claims that without it, he would not have gained the knowledge necessary to implement his entrepreneurial and social innovations. One of his fondest memories from his undergraduate days was being encouraged to explore Big Ideas—the ones that could change the world. When he would question the status quo, he was told, "Why not!" Tom's program at McMaster, which combined management with engineering and physics, convinced him that innovation occurs at the boundary between disciplines. Crossing from one discipline to another leads one to think outside of the box, combining information from different sources to explore new possibilities and apply different methodologies to create new solutions to problems.

His first job after graduation was working at McMaster's linear accelerator as a nuclear engineer, where he also taught fundamental principles of material sciences. However, after the Three Mile Island accident in 1979, which resulted in the partial meltdown of a nuclear reactor, Tom decided to switch majors to electrical engineering and studied semiconductors for his master's degree at the University of Toronto. He started working on patents and innovations in the field of semiconductors, which would prove useful in his later work in building networks and scanners. As a graduate student, he worked in a co-op program where he had the opportunity to see firsthand the possible applications of theoretical knowledge. He reflects that over the course of his education, he benefited from "a decade of tutelage from the finest minds in the field."

Tom also found an innovative way to earn two master's degrees at the same time. While working on his master's in electrical engineering at the University of Toronto, he asked if he might simultaneously enroll in an MBA. However, there was a rule that one could not enroll in two degrees at once at the University of Toronto. So instead, he registered at York University, also located in Toronto, for the MBA, completing both degrees at the same time, but in separate institutions. Simultaneously, he contributed to one of Waterloo's first spin-offs, which are companies based on the research being done at a university. Tom noted that one academic building housed tens of thousands of paper files, and it was difficult for staff to find anything. At the same time, the head of the computer science department, Dr. Savvas Chamberlain, was working on a hyper-tech search engine that could retrieve information that had been digitized. Most people did not realize the solution to their problem was so close to hand. However, in 1980, Chamberlain created DALSA Corp, now Teledyne, a leading company in the field of high performance, digital imaging, and semiconductors. Having collaborated with Chamberlain and shared a patent with him, Tom declares, "Sometimes opportunities for innovation are organic." Work in digital imaging took Tom to OpenText, which went from being a start-up to the first publicly traded internet company in Canada and the third on the Nasdaq exchange.

OpenText, the largest enterprise software company in Canadian history, named Tom as its first CEO in 1994 and chair of the board in 1998 and executive chair of the board and chief strategy officer in 2005, a role he filled until 2013. The company created the first internet-based document management system and the earliest versions of workflow, portals, and social networking software.

OpenText's goal is to digitally enable the world. By doing so, its work reduces the need for paper and results in the preservation of 1 percent of the world's trees. Tom extrapolates that its software is now used by half the planet, which means that trillions of trees have been saved while enabling people to communicate with one another around the world. Tom acknowledges that this realization was "a rare, humbling experience." However, before achieving this goal, there were many pieces of the puzzle to invent and assemble. For example, scanners had to be perfected and improved. At first they could only deal with machine-readable texts, but today they can deal with handwriting. They had to be both cost-effective and capable of detecting and uploading many forms of data. This meant that search engines had to be created to retrieve the information along with management systems to organize masses of information in a usable form.

The history of OpenText is a journey across Canada and through history. For example, while Tom and his team were improving the quality and attributes of scanners, they stopped, pausing their work to build a scanner specifically for use with the Mars Exploration Rover and the Robotic Arm so that it could record and transmit images from space. Tom recalls that when he and his colleagues actually saw the rover in geosynchronous orbit, they "felt as if they really did achieve something on a planetary level."

Another seminal moment in OpenText's success was in 1995 when Tom joined Jerry Yang, the co-founder and former CEO of Yahoo!, in a press conference to announce that OpenText would be providing the search engine for Yahoo!. The same engine was later adopted by IBM and Netscape. Being contracted by several of the giant international firms to purchase its technology was a very significant achievement for a small company in Canada. It also confirms that the engineering at OpenText was both ahead of its time and of great value.

"Pivotal moments like these are usually preceded by numerous nights when you're seized by a problem and cannot sleep as your mind tries to solve it. Sometimes you may wake up at two a.m. and write the solution down." Tom says that concentrating deeply on a problem is "both a blessing and a curse. Sometimes going for a walk while pondering helps." However, he cautions with a note of humour, "one has to try not to walk into telephone poles!" He thinks it is possible to train our minds to be innovative and find creative solutions to issues. We can teach ourselves to learn with structure and methodology. We need to learn the "nomenclature of a profession in order to ask questions." By this, Tom is referring to the logical structure as well as the vocabulary used in the specific field, both of which can be acquired through thoughtful research and education. Tom elegantly describes education as "a conversation between generations." Additionally, collaborating with people from different contextual and cultural situations may lead to new approaches to solving a problem. Tom believes that we best profit from having different approaches and viewpoints, especially when dealing with global issues. In this way, diversity and inclusion are key to innovation.

Today, Tom's work extends far beyond OpenText. He is currently chair of the World Wide Web Foundation, an international not-for-profit organization located in the US, advocating for a "free and open web." He has served on many boards and was notably chair of NSERC, chancellor of the University of Waterloo, and one of the founders of Communitech in Waterloo.

The next problem Tom wants to solve is the future management of robots and machines. With Major General David Fraser, who directed forces in Afghanistan, and Mark J. Barrenechea, Tom wrote a book titled *The Anticipant Organization: New Rules for Leading in Digital Society from the Boardroom and the Battlefield.* He says, "Business leaders need to understand the potentials and pitfalls of automation. E-government and e-business are happening today, and we need to harness digital potential positively." The alternative is aptly expressed in Tom's book with Mark J. Barrenechea titled *Digital: Disrupt or Die*. We have seen the power of computing when, for example, on Wall Street, computers archiving for derivative trading were able to process the information in nanoseconds—before humans could catch up. We must determine how to handle this extraordinary power in the service of humankind.

Tom's advice is to "keep at your dream. Do not fear failure. Entrepreneurs, researchers, and inventors all experience failure. Embrace it and consider it part of the journey." Tom's message is one of hope, and his story stands as a testament to the endless possibilities that lie before us if we are prepared to dream, work hard, and take risks like Tom Jenkins, an unrivalled expert in innovation.

4

DAVE SMARDON

Innovation in Agribusiness

Guelph, ON

Dave Smardon has a clear goal and a bold plan to make Canada's agricultural business sector more competitive and one of the best in the world. It is perhaps no surprise that he is CEO of Bioenterprise Corporation, "North America's leading accelerator for agricultural technology-based business," which may include finding alternate solutions to chemical fertilization, the use of pesticides, or weather-related issues. He is also CEO and managing partner of Bioenterprise Capital Ventures, a venture capital fund that invests in "emerging and innovative global agricultural firms" specializing, for example, in food and beverage processing or aquafarming. Led by a team of diverse and creative experts, Bioenterprise has over three hundred partners and members across the country, making up what is called Canada's Food & Agri-Tech Engine. The two arms of Bioenterprise have shifted somewhat from an emphasis on the bioeconomy and finding bio-based energy sources to a greater concentration on food security, sustainability, and climate change. Much of the company's success is in large part due to Dave's thoughtful leadership, passion, innovative ideas, and personal investments.

Dave was born in Toronto in 1954. Growing up, his family often moved due to his father's work, leading Dave to attend eleven different schools in thirteen years. He says this experience taught him not to be anxious about change and to be a good communicator. He also liked reading and watching science fiction series; whenever he saw something futuristic, he always thought, "Wouldn't it be great if we had that!" This curiosity and forward-thinking would certainly serve him well later in life.

In 1972, Dave graduated from the University of Toronto with an honours degree in economics. He went on to work for Texas Instruments and Unisys Corp, where he began to be inspired about the possibilities of innovation and technology. When he left to join Apple Computer, Inc. in 1996, he was impressed by their unique philosophy that encouraged entrepreneurship and mentorship. For example, after five years of employment, employees were granted a sabbatical to do whatever they wanted; they were even encouraged to create their own start-ups. To this day, Dave gives much credit to Norman Kirkpatrick, his mentor at Apple who encouraged him to be entrepreneurial and create start-up companies. He also recalls the impact of Apple's annual conferences where Steve Jobs would dramatically unveil impressive new technology that may have seemed like science fiction at the time, but would be adopted very quickly.

With three colleagues from Apple, Dave started his first software company, which was later bought by Unisys. His second start-up, also in the area of software development, was purchased by a Japanese firm. Thanks to these successes he was able to turn his attention to work in venture capital and headed up an arm of Apple's venture capital group. He also created his own leading venture capital firm, Nibiru Capital Management.

Dave was invited to speak at Guelph University about innovation and how it helps drive competitiveness, particularly when it comes to creating start-ups. There, he met Larry Milligan, a visionary who was at that time Guelph's vice president for research. Larry spoke about the many innovations in the field of agriculture and how other countries were succeeding, while Canada was falling behind. For example, Bolivia and Peru were producing similar food crops to those grown in Canada and were improving their products' resilience to weather conditions and thus increasing their yields to the detriment of Canadian producers, who were not motivated to collaborate regionally or nationally and therefore could not compete effectively. Dave offers as an illustration an asparagus farmer in Southern Ontario who developed a conveyor belt sensor camera. This was their strategic advantage and sharing it would only help their competitors, so there was no incentive to collaborate. Another major problem was that the Canadian innovation community was siloed: universities, incubators, and accelerators across the country worked independently and were not very well informed about one another's work. Larry asked Dave to work with them at Guelph to solve this issue, and Dave, having been suitably impressed by the work already begun and inspired by what might be accomplished if a national network could be established, de-siloing the agricultural enterprise, enthusiastically accepted the invitation. In 2003, he joined the fledgling Bioenterprise, which had only begun in 2002, with the goal of bringing knowledge, experience, and interests together to support and encourage the growth of innovative, new, and emerging agricultural companies across Canada.

Dave was encouraged by the work that had already been accomplished. After all, the agricultural sector is based on the work of farmers, who are some of the most innovative people on earth. They have been improving on their work for centuries: When a machine does not work, they fix it. When a crop fails, they study the reasons and improve the conditions. They continue to create and advance technology to make farming more efficient. There have been significant evolutions in Canadian agriculture, such as the increase in container farms, where climate-controlled boxes allow farmers to raise food anywhere, anytime; farms in northern Saskatchewan that now grow mango trees; and tobacco farms in Southern Ontario that have been transformed to produce ginseng. What was lacking was coordination and support.

Dave thoughtfully did research and began by looking at other countries like Australia, Brazil, the United Kingdom, France, and the US, which had all found ways to have a collaborative, cooperative ecosystem. Then he consulted business and industry, learning how they engaged with the agricultural sector's innovation community. The general consensus was that it was extremely challenging to acquire knowledge about Canadian innovations or even figure out where to go to obtain such information, and that most people would appreciate a central, coordinating body. Next, Dave visited universities, where people loved the idea of a centralized system that could provide connections to business and industry and offer new research opportunities. He looked at surveys from the United Nations and Bloomberg that indicated that Canada was well positioned in research investment (number eight in the world)—but dropped to the twenty-first position when it came to moving innovation into the marketplace. Finally, there was the Barton Report, which appeared in 2017, offering a comprehensive series of recommendations by an advisory council for the federal government, highlighting the same issues and listing agriculture as a sector in which Canada should invest.

Having done the research and analyzed the problem, Dave set out to improve on what already existed and create a more efficient ecosystem, increase the scale of innovation, provide continuity and quality, and measure the impacts. As Dave put it, they needed to "think of Canada. Think bigger." Bioenterprise has expanded with strategically located hubs and brilliant results, reporting in 2022 a cumulative total of two thousand technologies launched, twenty-five hundred innovative agriculture-tech companies created, $268 million in revenue, and a return on investment of 200 to 1.

Dave has helped build a national ecosystem and create an environment that supports innovation and collaboration. For example, one sector of agriculture that has seen a great change is viticulture, or wine growing. Where once there was uniquely fierce competition among the regions of British Columbia, Ontario, Québec, and Nova Scotia, there is now considerable support and sharing of innovative ideas, processes, and technologies.

Dave says the key to success is to follow your passion, though he adds that sometimes it takes time to discover the job that will become your career. Until then, you need to have the inner confidence that will allow you to do what you want.

We all know that changing the way we think and how we do things is often difficult to accomplish, but Dave proves that it's possible. He has found a vocation bringing together farmers, researchers, and the marketplace. He ensures that great ideas find fertile soil, and he brings all parties together to implement new concepts and attract the investments needed to spur growth. Dave says that if he had not been at Guelph and been so inspired to help, none of this would have happened. He calls it serendipity. I call it the good fortune of agricultural business in Canada.

5

KAYLA ISABELLE

An Authentic Leader Innovating for Good

Ottawa, ON

The award-winning CEO of Startup Canada, Kayla Isabelle connects entrepreneurs to resources, provides support to 122,000 entrepreneurs annually, and finds innovative solutions to people's problems—no matter where they reside, how ready they are to launch their business, which groups they represent, or the type of business they plan to conduct.

Kayla was born in Toronto in 1992, and her family moved often because of her father's work. These experiences offered Kayla the opportunity to make new friends at each location and, as a result, she was not shy about speaking with strangers, a facility she uses daily in her work today.

Her parents also encouraged her to study, and Kayla's father wanted her to go to university in Canada. At the time, Kayla was attending high school in New Jersey and was unfamiliar with Canadian universities. After some research, she skipped a day of school and flew to Ottawa. On arrival, she took a taxi and began chatting with the driver, who said that when he had been a newcomer to Canada, Ottawa had given him so much that he wanted to "give back." So he gave Kayla a free tour of the city, which convinced her to pick Ottawa to attend university. In 2010, she went to Carleton University, where she studied communications and psychology. She began practicing her leadership and speaking skills as president of the university's communications society. Originally, she wanted to become a guidance counsellor, but, as she began to discover her talents in communications, she changed her plans and took a small detour: she still delivers advice, but to prospective entrepreneurs. She also worked with a communications firm completing projects on a wide variety of topics from clean tech to beauty products. She learned the art of connecting dots across disciplines and enjoyed the interdisciplinary space.

In 2019, she decided to expand her knowledge of the world and travelled to Africa, where she worked for the tourism board in Tanzania creating materials intended to attract businesses to the country. She then spent nine months travelling the world alone and discovering her independent identity while learning about entrepreneurship. Every three months she would meet a friend in England or Thailand, for example. The rest of the time she was completely alone and challenged herself to face solitude. She learned that she could rely on herself and that she could decide to be with other people, but that she did not need to be with others. She returned to Canada with renewed energy.

In 2020, when she applied for the position of CEO at Startup Canada, she realized that, at twenty-eight, she was still quite young, but believed that her age is her "superpower." When she was hired, she realized that she had a superb board from whom she could learn. They coached her, kept her accountable, and offered her solid business advice.

At Startup Canada, Kayla works with governments, major corporations like Google, other nonprofits, and accelerators to encourage innovation and provide advice and support for new enterprises. Kayla listens, teaches how to incorporate, how to overcome obstacles, measure utility, navigate the many programs available, find the one best suited to the proposed new enterprise, and seek help along the way for mental health, as that, too, may be needed in a stressful environment. Startup Canada offers "high touch moments" at scale and makes the journey less lonely. It offers both practical and emotional support, and can operate in a flexible, low-cost way. Kayla has also created a program to support women entrepreneurs and attributes the high success rate of women-owned businesses to the fact that they often reach out to agencies like Startup Canada for advice and to their communities for product validation. As proof of Kayla's success at Startup Canada, the number of inquiries annually received from developers of prospective businesses has increased from 60,000 or 70,000 to 122,000.

She had just begun her role as CEO when the pandemic hit, and so she became an expert at pivoting as well as leading. Rather than create new programs, she concentrated on helping entrepreneurs access existing support. This was innovative and efficient, setting her organization apart as a "connector and a supporter for nascent businesses in a time of crisis."

Kayla believes that the impacts of the pandemic on businesses are deep, and recovery will take quite a while. She notes that small businesses are not yet "out of danger." The marketplace has changed, and small businesses across the country are finding it difficult to be profitable. Kayla works with these entrepreneurs through investment readiness programs, helping develop business plans, while doing "good works" that contribute to the environment, health, and the social fabric of communities. As the world moves into a post-pandemic phase, Kayla believes this is the right moment for innovation. The times are challenging, but then innovation is never easy!

She says that the concept of a "eureka moment" does not apply to her. Rather, there are micro eureka moments when she sees the impact of her work, especially when it helps Indigenous or women entrepreneurs, and when social enterprises succeed at helping communities. She thinks immediately of people such as Jenn Harper, who founded the first Indigenous-owned beauty company, Cheekbone Beauty, which supports educational opportunities for Indigenous youth and creates a space where everyone can feel represented. Birch Bark Coffee Company, founded by Mark Marsolais-Nahwegahbow, similarly contributes part of its profits to clean water for Indigenous communities. Kayla also points to a company that makes soap and its founders Jacqueline Sofia and Noora Sharrab, who created Sitti Social Enterprise, which provides employment to women in refugee camps in the Middle East, helping them gain skills and needed revenue.

Kayla is very modest about her achievements, but she has received truly well-deserved recognition for her work, including the 2021 RevolutionHER Impact Award in Leadership and the 2021 Women of Inspiration Authentic Leader Award; she has also been named a 2023 Changemaker by the *Globe and Mail* and a 2022 Forty Under 40 by the *Ottawa Business Journal* and the Ottawa Board of Trade. Her commitment, energy, and enthusiasm are contagious, and it is not surprising that she has inspired many companies to move along the path to success.

Kayla's advice is that "one should not hold on to discouraging moments, such as a job one did not get. These are only micro-moments in life, and we should not waste energy on them, but rather move forward." Kayla is a truly admirable and engaging person who works with her heart as well as her mind. She appreciates the solitary nature of work, but also values communication and collaboration. She recognizes the particular need for social enterprises, whether in rural Canada or around the globe. She has learned to appreciate and nurture the work of others, and there is undoubtedly a successful future ahead for Kayla and the thousands of entrepreneurs she will inspire.

6

KEN DOYLE

Creating the Conditions for Success

Ottawa, ON

Ken Doyle, executive director of Tech-Access Canada, is both an innovator and a supporter of social and economic change. A brilliant scholar, he modestly attributes his success to on-the-job experience. He is a talented creator; he sees possibilities, makes connections, and brings people with potential together with proven business leaders. A wise and perspicacious judge of markets and the potential for development, he understands that success lies in the participation of underrepresented groups and designs programs to ensure this happens.

In 1982, Ken was born in modest circumstances in Sussex, New Brunswick, but his parents and grandparents instilled in him the value of education from a young age. It was understood that he would one day attend university and that they would provide the support for him to study. Of course, he played sports and video games; he also had a paper route, where collecting the fees taught him how to be less of an introvert.

In 1988, the province of New Brunswick decided to mark the upcoming new millennium: each elementary school would pick one student who would graduate from grade twelve in 2000 to be specially recognized on their graduation. At the age of six, Ken was selected as his school's representative and got to shake the premier's hand at a ceremony. However, he did not receive the special recognition on graduation, as his family left New Brunswick and moved to Ontario, where secondary school required an additional year (grade thirteen) at that time, and so Ken ended up graduating in 2001 without fanfare. He learned the difference a year can make in life, and that timing is everything.

During high school, Ken was placed in French immersion, a bilingual program, with an enriched curriculum. His only regret was that he could not take the auto mechanics course at the same time, for cars were his passion. He applied for after-school jobs to be able to afford a car, which would give him a measure of autonomy. When he was accepted into Carleton University in 2001 to study public affairs and policy management, he got a night job fueling planes at the airport. Since airports operate all day, every day of the year, and in all weather conditions, the job taught him a strong work ethic, effective problem-solving, and how to keep cool under pressure.

On September 11, 2001, Ken took the bus to university as usual. When he arrived, everyone was gathered around the television. The atmosphere was somber. The world had dramatically changed with the terrorist attacks on the World Trade Center in New York. That evening, Ken went for his usual night shift at the airport fuel storage facility, where the two one-million-litre fuel tanks were kept and where he worked alone. That was the first time that he thought about the risk to his personal safety his job involved. It required climbing to the top of huge metal tanks and manually validating the computer reading of their total remaining fuel. He remembers thinking about the Roman philosopher Boethius, who described life as a wheel of fortune that rose and fell, bringing people to wealth and power or to poverty and hunger with a turn of the wheel. He determined to do his utmost to turn that wheel of fortune in the right direction.

Ken realized that he had been very lucky not to have slipped and fallen from his precarious perch and decided that he no longer wanted to work at a dangerous, low-paying job where he had a great deal of responsibility and no authority. Ken redoubled his efforts both at university and work, determined to qualify for a career where he could make a more significant contribution by leading operations. He took on an additional job at the airport handling baggage, where he still may hold the Ottawa single-flight record for most baggage loaded into the rear compartment of a 737. He progressed to a job at the Shell Aerocentre, fueling and handling aircraft and making lifelong friends who span the country and the globe.

Upon graduation in 2005, he saw a poster announcing that Statistics Canada had a fast-track, eighteen-month program for economists. He applied and was offered the job, which would include training. At the same time, he also received a job offer from the Association of Canadian Public Polytechnic Institutes (ACPPI), which is now known as Polytechnics Canada, an advocacy group that aimed to increase funding for eight large urban colleges doing a considerable amount of applied research and offering bachelor's degrees as well as skilled trades training. He asked his father which job he should accept. His father said the association sounded interesting. It might offer Ken a niche, and if he did not like it or if it failed, he could always apply to work in the public service. Ken followed his father's advice and never turned back.

Well aware of the roadblocks in industry and society that prevented the apprenticeship and skilled trades programs from flourishing, including the costs that were both a challenge to small companies and a financial burden for the individuals, the length of the program and the fact that without significant growth, new apprentices were not required, Ken turned his attention to improving other programs at ACPPI. He was proud of his contribution to the Natural Sciences and Engineering Research Council of Canada's fund that was tailored to support colleges, resulting in increased support for applied research. He also engaged with the Social Sciences and Humanities Research Council of Canada (SSHRC) to promote interest in social innovation. He contributed to the creation of the Canada Apprentice Loan, which supports apprenticeships with interest-free loans. Ken worked alongside FedDev Ontario, a federal economic development agency for Southern Ontario whose goal was to develop the economy through diversification, scaling up business, creating networks, and supporting business needs. He applauded the FedDev Ontario pilot program, which reversed the habitual pattern of funding companies directly, encouraging them to engage instead in the Applied Research and Commercialization Initiative (ARCI) program, which gave vouchers to universities and colleges, which then had the task of locating and working with companies in order to be able to cash in the vouchers. In parallel, Ken was working toward his master's in public policy and administration with a concentration in innovation, science, and the environment at Carleton. He found it much more rewarding than his undergraduate experience, as he was able to draw from real-life work experience to amplify the course content.

Ken thoughtfully considers the best companies in which we should invest in order to improve Canada's economy. He notes that the average size of a company in Canada is eight employees, and 70 percent of them only have four employees. There are 1.1 million companies in Canada, but only 20,000 claim to be doing research (down 20 percent from ten years ago). Indeed, there are 990,000 small companies. There are very few companies in the middle or at the top of the scale. This is different from Europe, where they have many more midsized companies with more employees and a higher commitment to research and development (R&D). As a result, they have the capacity and capability to execute these activities in-house. The conclusion is evident. We need to expand the number of midsized and large companies and encourage them to invest in research. To accomplish this, we need to grow the small companies.

After ten years with ACPPI, Ken stepped down. He was proud of his accomplishments, but his heart was no longer there. He was asked to take on some consulting for Colleges and Institutes Canada, where he was special advisor to the vice president, government relations, and Canadian partnerships. Then he had the opportunity to become the inaugural executive director of Tech-Access Canada (originally Réseau TACCAT Network), where he has had "a wild journey since 2016."

Tech-Access Canada is a nonprofit that supports the pan-Canadian network of Canada's Technology Access Centres (TACs). TACs are specialized R&D centres affiliated exclusively with public colleges, which help small Canadian companies solve their business innovation challenges. The TAC model is based on the long-standing Centres collégiaux de transfert de technologie in Québec. The member TACs set up a modest two-year pilot program in 2016, with an ambitious mandate to share best practices, raise awareness of the centres, and respond to community needs. The centres expanded in number rapidly from twenty-five to sixty. The National Research Council–Industrial Research Assistance Program (NRC-IRAP) was the first to recognize the true potential of the network. IRAP was established to assist small and medium-sized businesses increase innovation capacity and take ideas to market. TAC worked with NRC-IRAP to provide support for developing and refining prototypes, validating new technologies, and getting objective advice on business innovation challenges. Together, they successfully extended the program across Canada.

Ken says that while NRC-IRAP helps companies that are pre-screened to be "Canada's next unicorns," it has a finite budget and many companies do not make the cut, including a number established by recent immigrants and women. These businesses were generally considered too small, too "niche," or too recently established. He wanted to support this cohort and noted that the 2021 Canadian federal budget mentioned his program and funding to "encourage the participation of under-represented groups." This fiscal support meant that Ken could create a program to support and encourage these groups.

Ken asked his member TACs to seek out the companies that had been rejected or were on the sidelines and request them to apply again. In the first year of the improved program, 38 percent of clients voluntarily self-identified as being majority-owned by members of underrepresented groups. With strong support from eCampusOntario's leadership, Ken started an Inclusive Innovation pilot program that identified twenty-two projects, rapidly awarding $5,000 in innovation support to each. The success was incredible. For example, in Saint-Jérôme, Québec, there was an area of pristine lakes where gas-powered boats were banned to reduce both noise and water pollution. However, people still wanted to water-ski, and so a company wondered if it could retrofit a traditional water-skiing boat with an electric power train that would be as powerful as a gas motor, that could last as long as a tank of gas, and that could be charged overnight from the dock. Through a collaboration with the Innovative Vehicle Institute, the boat the company described was designed and ended up being even more powerful than a gas-powered one. The TACs gave the intellectual property to the company for commercial exploitation and, in return, its students received innovation skills training via the R&D project teams.

There is a 93 percent successful match rate between the TACs and the businesses that contact Ken with their innovation challenges through the TAC Jump Ball program, which connects businesses with questions and experts, who are prepared to assist and advise. Ken is by no means slowing down, though: he has a list of thirteen public policy pilot ideas that are ready to launch and address remaining challenges, such as the talent shortage. Industry wants "plug and play talent." They want to hire employees who are completely qualified and ready to perform from the first day on the job. This excludes people who may not yet have acquired the right skills and need additional training. As people retire, new talent is sorely needed. Ken provides a way to bring people and needs together and to provide the training or networking required for success. Ken's pilot ideas build on market failures within existing programs and propose something unique and solution-oriented. He is driven to develop these pilot programs to allow dreamers, researchers, investors, and governments to collaborate and put Canada back on the map as an innovation nation.

Ken says one should "be humble and this will guarantee that you will learn something," and that "failure is an acceptable outcome of innovation. Not all the balls you hit will result in home runs." Ken thinks of all the big companies that have failed in Canada and says it is unfortunate that we forget about them so quickly, despite their previous successes. For example, he would like to bring together the heads of Corel, Nortel, and BlackBerry, and ask them to talk through their companies' rise and fall. If they had a second chance, what might they have done differently? Ken is always trying to learn and think of new ideas: he says that inspiration can strike anywhere, and always carries a notebook to jot down ideas he may suddenly have.

Ken's life is an example of innovation: he wisely chose his path and works diligently to pursue his goals. A social innovator, he has always analyzed problems and recognized areas where the systemic challenges were too great for a single individual to solve—and has encouraged collaboration to correct problems, and he has personally advised and assisted aspiring entrepreneurs. Together, people can turn that wheel of fortune and create a better future by enabling small entrepreneurs to grow to the point where they, in turn, will help others by investing in research that will lead to winning ideas. With each of his new and creative initiatives, Ken engineers successful, innovative, social, and economic change across the country.

PART TWO

Saving Lives with Innovations in Medicine and Medical Devices

These researchers are pioneers in their fields. They are cardiologists, neuroscientists, geneticists, medical engineers, and molecular biologists who are passionate about their work, which they often characterize as the quest to open a "final frontier" of knowledge. Once the complexities of the nervous system, the brain, the heart, or the genetic makeup of viruses are understood, effective vaccines or treatments that will directly target specific viruses can be developed. This means less damage to surrounding cells in treatment, and it means crossing another frontier that will lead to new possibilities.

When surgery is required, these researchers are exploring ways to make it more precise, and to ensure a more rapid recovery. Innovations in equipment are also a source of pride for Canada. A robotic arm, created in collaboration with MacDonald, Dettwiler & Associates, the company responsible for the Canadarm used in space, is now employed in operating rooms around the world. The thoughtful technologies and processes that increase patient comfort during the diagnosis and treatment of cancer, and the more precise and effectively tailored creation and delivery of drugs or radiation, offer examples of contributions by Canadian researchers whose inventions are considered the new global standard.

Concern about the overuse of antibiotics has driven the discovery of the antimicrobial qualities of naturally existing corals or plants that will enable the creation of new products to eliminate infections. This century will, I am sure, be remembered in medical history for the incredible progress and new discoveries that have been made. These pioneers reflect on their "eureka moments" and the hard work and failures along the way. They share their hopes for the future, and the passion that drives them. Their stories are truly inspiring.

7

MICHAEL HOUGHTON

A Noble Quest for Safe Transfusions and a Hepatitis C Vaccine

Edmonton, AB

Nobel laureate Sir Michael Houghton grew up in a working-class family in post–World War II London, England. His father wanted him to earn a good living by going into business or becoming an accountant. However, Michael declared that his brother was the mathematician in the family and he himself would prefer to study science. Michael's passion for biology would lead to his receiving the Nobel Prize in Physiology or Medicine in 2020 and becoming knighted in 2021.

It all started with a science program on BBC Radio that aired on Sunday mornings. The show introduced the fourteen-year-old Michael to American biologist James Watson, English physicist Francis Crick, British chemist Rosalind Porter, and British biophysicist Maurice Wilkins, who uncovered the mysteries of DNA. Watson and Crick received the Nobel Prize in 1962 for discovering the double-helix structure of DNA, and demonstrating how it replicates and is coded with hereditary information. The show captured Michael's imagination and inspired him with a desire to study biology. Unfortunately, biology was not available at his high school, so he studied math, chemistry, and physics instead, and nurtured his interest in the biosciences by reading about the life of French chemist and microbiologist Louis

Pasteur, who pursued his scientific research in vaccines and pasteurization after he lost his own daughter to infections.

In 1966 Michael attended the recently established East Anglia University. He was pleased to learn that it offered courses in biology, but he was disappointed to discover that they were only in general biology and not in DNA and RNA or the emerging study of molecular biology.

On graduation, Michael could not immediately pursue his PhD at King's College London because he had not been able to take the required prerequisite specialized courses in molecular biology. So in 1972, he applied for a job working as a research assistant with G. D. Searle and Company, an American pharmaceutical company and a subsidiary of Pfizer, located in London. There, he and his colleagues, a fine international team of researchers, linked hepatitis C with liver cancer. Not even a year later, he was admitted to King's College, where he completed a PhD in biochemistry in 1977. From then on, Michael's work has focused largely on hepatitis C, which is a virus in the same SARS family as COVID. While early symptoms of hepatitis may include migraines, general fatigue, and discolored urine, hepatitis C may go unnoticed for some time, affecting the liver, and result in cirrhosis or cancer. It was only in 1978 that co–Nobel laureate Harvey J. Alter showed that the disease was transmissible via the transfusion of donated blood.

Upon completion of his studies, Michael worked for several pharmaceutical companies in England and the US, over the course of more than a decade. With his colleagues at Chiron Corporation—Qui-Lim Choo, from Singapore, who co-discovered hepatitis D and worked on the blood test that could detect hepatitis C; George Kuo, a Taiwanese graduate of the Albert Einstein College of Medicine; as well as Daniel W. Bradley, an American geneticist expert on the chimpanzee model—Michael discovered the hepatitis genome in 1986 and hepatitis C in 1989. These discoveries took years of work. What Michael and his team did differently from previous researchers who had worked to identify the disease and confirmed that it was transmittable was to apply molecular biological approaches such as gene cloning, recombinant DNA, and allied technologies to isolating the gene itself in order to create a vaccine to prevent the disease from spreading. They tried thirty or forty different approaches to cloning the hepatitis C gene until one finally succeeded. The limitations of the technologies that were available at the time forced Michael and his colleagues to push their machines to the limit—to the very "edge of sensitivity"—in their ability to make accurate measurements of extremely minute samples of genetic matter. They ended up using a novel screening technique to isolate and clone the hepatitis C virus and create a vaccine in the lab. In 1994, they published a paper with the National Academy of Sciences, showing theoretically that the vaccine for hepatitis C could work.

At that point, the vaccine had to be tested and the only animal whose DNA was sufficiently close to that of the human was the chimpanzee, an endangered species that could not be used in testing. The scientists would have to find an innovative way to solve the problem. They tried to grow the hepatitis virus in cell culture in the lab. To accomplish this task, they had first to find an antibody that would make the virus less infectious. The virus also proved very hard to grow. It needed special host factors found only in liver cells. When the team transformed cell lines, they lost the ability to provide the host the factors required to successfully replicate the specific virus, which instead mutated into different strains, meaning that the hepatitis C vaccine could not be tested.

Michael and his colleagues tried thousands of strains of the virus before they finally found one that could grow in the cell culture. This strain came from Japan. They knew that if they could make a version of the virus in that genome, using the appropriate culture, they could create an effective vaccine against the hepatitis C virus. Michael's team was also attempting to create a diagnostic test that could be used on blood samples to confirm presence of the virus in people, and to create antiviral medications and remedial treatments.

In 2003 they paused their efforts to develop a vaccine for SARS-CoV-1, a severe acute respiratory coronavirus that had reached epidemic proportions, infecting some eight thousand people in thirty countries. This virus was carried through saliva droplets in the air and on surfaces and resulted in many deaths. However, the virus died out on its own, and the vaccine developed in Michael's lab was never manufactured. Had it been produced, Michael and others who also worked on the SARS-CoV-1 vaccine assert it would have been effective against the COVID-19 virus.

In 2004, Michael and his colleagues resumed their work on a vaccine for hepatitis C, while other researchers in labs around the world developed diagnostic reagents to detect the virus in blood supplies to reduce the risk of exposure from one in three million to one in two million. Antibody testing likely has prevented forty thousand new infections each year.

Michael came to Canada in 2010 as a Canada Excellence Research Chair, an award established to attract outstanding scholars and researchers to Canada. Then, in 2013, his group was the first to show that the hepatitis C vaccine can be effective against all strains of the virus and the first to show that in human trials the vaccine produced antibodies that could neutralize the virus.

Today in his lab at the University of Alberta, along with leading virologist John Law, Michael has developed a vaccine with two varieties: a protein version and an RNA version. The first is the traditionally formulated vaccine, and both are in clinical trials. Michael hopes that the vaccine will be produced, potentially saving four hundred thousand lives each year. The incidence of hepatitis C has been increasing in part due to the opioid pandemic. It is extremely expensive to treat patients with antivirals, which cost $20,000 per patient. In Canada alone, there are some 250,000 carriers. This means the vaccine has the potential to not only save lives, but to save $5 billion in drug costs alone.

When Michael looks to the future, he sees his next challenge as producing one vaccine for both hepatitis C and HIV. This is important, as those who might come in contact with blood or bodily fluids such as health-care workers, emergency responders, drug users, or those having sexual contact with an infected person would at once be susceptible to either or both diseases. The vaccine could also increase the life expectancy of those receiving antiretroviral therapy for HIV. He cites promising news of an important finding. Recently, in Ghent, Belgium, a new strain of the hepatitis virus has been found that elicits antibodies that neutralize most other strains. His dream would be to find one vaccine effective against all strains of SARS-related viruses, which include COVID, hepatitis, and HIV.

Michael attributes much of his success to professors and colleagues. He enjoyed the generous mentorship of Norman Carey, a virologist from the University of Cambridge in England, and interacted with Richard Palmiter, a cell biologist from the University of Washington, to work on virology, immunology, and cell botany. He is prompt to state that he won the Nobel Prize with Charles M. Rice, the American scientist who first identified hepatitis C, and Harvey J. Alter from California Tech, who discovered the way the disease was transmitted, while Michael himself cloned and isolated the virus, subsequently creating the vaccine. In addition, Michael was perhaps the first person to turn down the prestigious Gairdner Award in Canada unless his fellow Nobel laureates could officially be honoured with him. He strongly supports collegial discussions and sharing both honours and ideas, which, he says, "occur while interacting with other scientists and usually not at formal meetings. It is when you are having a cup of coffee or tea and you are talking about a baseball game or a cricket match, and then return to work that you have a brilliant idea," he says. "In my experience, those are the best and most fertile discussions—when two or three minds connect and stimulate each other. That is when you come up with a new avenue to pursue."

The world is fortunate that Michael Houghton, now Sir Michael, did not follow his father's career advice; nor did he despair when he was sent off to seek his fortune with best wishes and a new suit and tie; and he did not give up on his dream to work on DNA-RNA. Instead, he used his passion, curiosity, and thirst for knowledge to find and build a strong team of like-minded colleagues who were equally determined to succeed. He also applied a new approach in the search for a vaccine. When more traditional ways of testing were not available, he invented others. When new technology became available, he was an early adopter.

Dr. Houghton's family have noted that for many years he has worked late nearly every evening and was always in his lab on Saturday. Thankfully, he is still following his passion.

8

MATTHEW MILLER

Creating New Vaccines and Innovative Ways to Deliver Them

Hamilton, ON

When Matthew Miller would head to school in his hometown of Belleville, Ontario, his parents would say to him, "Remember, Matthew, you represent the family."

And in this way they also encouraged him to study. He received a well-rounded education, with a good balance of science and social science. As a child he liked history and thought of becoming a writer. But when he was in grades eleven and twelve, he was "bitten by the science bug" and determined to specialize in biotechnology in university.

He worked steadily each summer, and his very first job was stocking the shelves at a local pharmacy. He later delivered surgical equipment and supplies to local hospitals, worked in auto parts in a manufacturing facility for vehicles, and did supply teaching. While the teaching may have helped pave the way to his career, the other jobs served not only to help finance his education but to help him understand supply chains as well.

In 2003, Matthew enrolled in a newly minted program at Western University in London, Ontario, a bachelor's degree in medical sciences that allowed him to tailor his own specialization. He chose microbiology and immunology. In his second year he was "absolutely taken" by a microbiology course, which he "consumed voraciously," and in his fourth year he completed an honours thesis that was worth half his final year. He spent the entire semester in the molecular biology lab working on the human cytomegalovirus. He discovered that he loved research, and the lab environment was like a close-knit family, so he decided to continue on to do graduate work. After receiving positive encouragement from his professors, as well as early awards and recognition, his work in virology seamlessly became his career trajectory.

From 2007 to 2011, Matthew worked on the "Red Queen hypothesis" for his master's and PhD at Western. Originating in evolutionary biology in 1973, the hypothesis stated that species must constantly adapt, evolve, and proliferate in order to survive against ever-evolving opposing species. Matthew says that pathogens and hosts are always in a race against each other, but they also need each other to survive, so they are all working as fast as they can to evolve. He also notes that some viruses program the human cells in which they are located in order to self-replicate, thereby assuring the continued survival of the viral cells in the body. Herpes viruses are a good example—they stay with you all your life. Chicken pox may reappear as cold sores or as shingles years after the first infection. These "hit and run" viruses remain dormant in a host over long periods of time and periodically reactivate, adding another layer of complexity in developing a vaccine. A vaccine for HIV, for example, must work throughout the infected individual's lifetime, even through the periods of a virus's dormancy.

In 2011, Matthew went to the Icahn School of Medicine at Mount Sinai in New York and worked with renowned American microbiologist Peter Palese, who built the first genetic maps for influenza strains A, B, and C. During the SARS epidemic of 2002 to 2004, Palese's lab had discovered that while the antibodies produced by people who had been previously infected with the virus could protect against many strains, revaccination would still be required after some time. The goal was to instill long-term immunity in the population. The Icahn lab had developed the first-ever universal vaccine candidate, and it had just finished phase one testing, which did well eliciting responses at the end point. The universal vaccine was not developed further, however, because the SARS virus had died out on its own. This then was the subject of Matthew's work while in New York. He wondered how antibodies were formed after vaccination and why they diminished in effectiveness over time. He learned a great deal that would become useful when the COVID-19 pandemic occurred.

Matthew reminds us that the 1918 Spanish flu killed approximately 17.4 million people, or 3 to 5 percent of the world's population, which is the equivalent of 360 million people today, or approximately the total population of the United States and Canada put together. The COVID-19 pandemic caused over 6 million deaths, or 1 percent of the world's population. Despite incredible medical advances, the world was not well prepared to prevent the pandemic.

In 2014, Matthew returned to Canada to McMaster University, drawn by Karen Mossman, who is the vice president of research and innovation at the university and who had served as the external examiner on his thesis in New York. Then the COVID-19 virus hit in December 2019, spreading throughout the world in just a few months. Matthew shifted his target from the SARS virus to COVID. He focused on creating a vaccine that could be administered directly to the lungs, a unique inhalable form of vaccine, since that is where COVID first attacks, thus providing more rapid and effective treatment than an injection in the muscle tissue of the arm. In addition, a vaccine he had developed against SARS-CoV-2 also proved effective against variants of concern.

Matthew consulted engineers about aerosol dynamics, and together they manufactured a vial with a spray inhaler that could be dispensed at a pharmacy. The inhaler, which has successfully undergone clinical trials, offers a more effective and rapid way to vaccinate the population.

In 2022, Matthew became the director of the Michael G. DeGroote Institute for Infectious Disease Research at McMaster University. The institute has already produced a novel vaccine for tuberculosis, as well as innovative methods for cancer treatment, preventing and managing lung disease, and treating food allergies. The institute's success can be attributed to cross-disciplinary research between biological anthropologists and social scientists who work in close collaboration with medical researchers. Together, they are developing an understanding of the intersection of viral infections and the characteristics of the individual human genome, particularly in the case of autoimmune diseases like ALS (amyotrophic lateral sclerosis). They are learning that many of these diseases are multi-genetic and that environmental factors may also play a role. Matthew wants to find the cause of such illnesses and then determine ways to prevent them through vaccines. He works on influenza, the coronavirus, vaccines, antivirals, neurodegenerative diseases, and B cells. Matthew is currently working on a class of well-known antiviral drugs to treat seasonal influenza, protecting against all strains of the flu, preventing both flu and another pandemic.

Matthew tells his students to "do what they love."

"You cannot be happy in life if you are not fulfilled in work," he says. "Too many people are driven to careers for financial rewards. Who I am is what I do, and I love it. And my love for my work is more important than anything money can buy."

9

ANNA BLAKNEY

The Future: From Self-Amplifying RNA to Cool Communications

Vancouver, BC

Open your TikTok app and look up Anna Blakney. You will be the 270,001st person to meet Anna and some of her colleagues, see her lab, and admire the commitment and energy of this extraordinary and personable early-career researcher.

Let us start at the beginning, in 2002, in Colorado, where Anna was inspired by her middle school math teacher, Ms. Hogue. Ms. Hogue encouraged Anna to join the after-school math club, where she gained confidence in her own ability, even though math wasn't exactly "cool" in grade eight. Then, when she was in high school, she attended a summer engineering camp, where the university faculty did their best to persuade students that engineering was fun with activities like the (in)famous egg drop, where science students across North America traditionally release various objects, including eggs and melons, from the roof of a building and record the timing, weight, and arc of the falling object, as well as, in certain cases, its ability to withstand impact due to a student-designed protective wrapping. For campuses with access to water, another activity involves the construction of rafts out of "funky" materials. Anna decided to follow her father and grandfather and become an engineer. However, she broke with the family tradition of chemical engineering, opting for biological applications instead.

In 2008, during her first year at the University of Colorado at Boulder, she was given the opportunity to work on 3D synthetic hydrogels in a tissue engineering lab directed by chemical and biological engineer Dr. Stephanie Bryant, winner of the university's 2007 New Inventor of the Year award for "encouraging the next generation of scientists to push boundaries and bridge disciplines." Anna worked there for all four years of her undergraduate program, growing cell cultures in three dimensions using polymers. These would be used to replace animal models in research. Prior to this experience, Anna had no idea what research was or of how you qualified to become a researcher. Dr. Bryant met with each of her students, one-on-one, every week, and she encouraged Anna to go on to graduate school.

Prior to embarking on her graduate studies, Anna worked one summer in a Big Pharma lab, which, compared to her work in the university lab, was mission-focused and lacked the freedom to explore, which Anna realized was important to her. In 2012, she went to the University of Washington in Seattle, where she completed a PhD in bioengineering, working to develop a new drug that could be used safely for pregnant patients who presented symptoms of HIV. As part of her doctoral program, in 2014 she spent six months doing research and working in the labs at the University of Cape Town. There, she studied how the tuberculosis (TB) vaccine (BCG) affected the susceptibility to HIV in children.

This was Anna's first experience abroad, and while she was impressed by the beauty of Cape Town, she was also struck by the disparity between the "haves" and the "have-nots." It was also the first time she had to worry about her personal safety; she could not go out in Cape Town after dark, and she had a panic button in her apartment in case of a home invasion (which she fortunately never had occasion to use). She realized how much she had taken for granted in her life: the freedom to go out, to attend a good school, to have free and rapid access to publications, to work in the most modern laboratories. The experience moved Anna to think about focusing her future research on projects that would assist women and underserved populations. While in South Africa, she also met her future postdoc supervisor, Dr. Robin Shattock, a renowned immunologist, and head of the Mucosal Infection and Immunity labs at Imperial College in London, England.

In 2016, she applied for, and received, a fellowship from the Whitaker Foundation in Arlington, Virginia, which supports bioengineering and medical research, and requires students to do postdoctoral studies abroad. That same year, she went on to work in a lab at Imperial College that was jointly directed by Shattock and Dr. Molly Stevens, a highly regarded specialist in biomedical materials and regenerative medicine. She was there for four years working on self-amplifying RNA for vaccines. Also termed "self-replicating," once implanted, the saRNA reproduces itself in cells, resulting in a higher expression of the proteins that will fight off the virus. It also means that a lower initial dose of RNA is required. It was a smaller field, but Anna felt the idea was "pretty wild, high risk, but interesting and cool." They had progressed to the point where clinical trials were planned, when the pandemic hit, and they pivoted to work on a vaccine for COVID-19.

In 2020, Anna was recruited by the University of British Columbia to work in the Michael Smith Laboratories, which combine biotechnology and chemistry and also foster interdisciplinary research and study. Anna's team includes plant biologists, specialists in RNA vaccines, physicists, and a person who handles biotech relations and outreach in science communications. Anna was always fascinated by the challenge of explaining scientific concepts to the public. During her graduate studies, she had the opportunity to do science outreach at the university and at the science museum. She had always enjoyed watching Bill Nye, the "science guy," on television, so she accepted an invitation to do a Reddit "Ask Me Anything." This program invites experts to respond to questions posed by the public. That fall, she and a colleague agreed to speak about the COVID-19 vaccine. The show was so popular, it made it to the Reddit front page. Shortly thereafter, Anna received a call from Team Halo, an organization associated with Verified, a United Nations initiative to deliver fact-based communications on global health to the general public. Working with the Vaccine Confidence Project at the London School of Hygiene & Tropical Medicine, Team Halo wanted scientists to participate in the project by going on TikTok.

Anna had never been on TikTok before, so she had to learn how to educate people about vaccines using a media that was foreign to her. She thought that, at the same time, she could show the public that scientists are normal people who have a sense of humour. Since her first video was uploaded in 2020, she has been creating four to seven videos every week and has garnered 270,000 regular followers.

Innovation means seeking new ways to solve problems. In this case, the general issue was to communicate scientific information to the public, whose trust in science was not strong. Anna meets viewers in their homes and explains science using simple terms.

Meanwhile, she continues her scientific research. Early on, she believed self-amplifying RNA (one type of mRNA, which is messenger ribonucleic acid, single-stranded molecules that carry a genetic code) could be used to create vaccines and was an early explorer in a field that now has much momentum. Around the world researchers are applying the knowledge that mRNA can be used to trigger the human body into creating antibodies, and then the vaccine itself will disappear, leaving only the antibodies. She was involved in the first clinical trial of a self-amplifying RNA vaccine. This innovation allows a more rapid production of a vaccine in the laboratory without using poultry, for example. One lab can sequence the RNA code and simply email it anywhere in the world. The vaccine also requires a lower dose of RNA, as it replicates itself in the human body.

Anna will be looking at different viruses to determine which combinations of RNA are effective in specific cells and notes that the application of machine learning will speed up the process. She offers the example of a vaccine for chlamydia, which can cause infertility. It was developed in Dr. Robin Shattock's lab, and is currently under trial.

Anna was also responsible for another innovation, which was to collect the excess human skin after plastic surgery and use it in her lab for testing instead of disposal. As a result, she gets readings that can be even more effective than those using mice. She started this process as a postdoc, not knowing if it would work. It has been extremely useful and has sped up preclinical trial work and possibly offers an eventual alternative to animal testing.

When asked what has inspired her, Anna points to her professors, who gave so much to her and to other students. She feels responsible to do the same. She is passionate about her research, about finding treatments that could be accessible to women living in poverty, and about saving lives. She points to the extraordinary opportunities she has had to study and do research, and the fact that, in other nations in the world, this simply would not have been possible. She also recognizes her parents, who have always been so supportive and who said, "Do whatever you want. But be good at it!" It is more than evident that she followed their advice.

10

MONA NEMER

A Scientist Leading Science

Ottawa, ON

A highly impressive scientist whose discoveries have saved many lives, Mona Nemer takes pleasure in recognizing the extraordinary accomplishments of Canadian researchers and inviting us to celebrate the innovation that flourishes in all fields and in every corner of our country. She has received many honours including the Order of Canada, fellowship in the Royal Society and the Chemical Institute of Canada, as well as knighthood in the Ordre national du Québec and in France's National Order of the Legion of Honor.

Born in Lebanon in 1957, Mona grew up at a time when there were separate schools for girls and boys. She liked mathematics and science, but by the age of fourteen she knew that she would never even have the opportunity to continue her studies in these subjects, as they were not offered in the school for girls. Consequently, she launched a petition against this policy, which resulted in making the science stream a part of the curriculum at the girls' school. Mona created the context that would open a world of opportunity for girls. She was already displaying remarkable innovation—and at such a young age!

In 1975, Mona enrolled at the American University of Beirut. The effects of the civil war in Lebanon (1975–1990), which was an armed conflict that caused 120,000 deaths and the emigration of a million citizens, made it difficult to study. She had to sleep in the basement of an academic building because the country was being bombed and some of the bombs even exploded on campus. She petitioned her parents to permit her to go to the United States to continue her studies in English, her third language after Arabic and French. Her parents agreed, but asked her to select a university in a state where they had relatives. That meant Florida or Kansas. She chose the latter and went to Wichita State University. At first, Mona felt as if the charming little city she had landed in was located on another planet, and the local population, unused to international students, regarded her as if she had just come from Mars. She dedicated herself to her studies and took as many courses as possible so she could graduate early. During that time, she made friends who gave her lessons in idiomatic English, and she came to appreciate the difference between undergraduate studies in the US and Canada. In Kansas, the curriculum followed the classic French course of study, which allowed for a wide range of subjects, and she took minors in French, mathematics, and chemistry with required courses in other disciplines like history and economics. The Canadian program, which she would later discover, put more emphasis on a student's major, but as a result was less well-rounded.

On graduating from Wichita State, Mona was offered a scholarship at the University of Michigan. Over the summer, she decided to go on a road trip with friends who were visiting from Lebanon. They went to Detroit, Toronto, and Montréal, where Mona fell in love with the city. After just two days, she decided to go to McGill University. Despite having applied late, she was accepted in a master's program and ended up doing a PhD in chemistry. She was interested in the development of drugs and was accepted in a molecular biology lab working on the chemical synthesis of RNA, where she discovered a true passion for the subject and realized that she wanted to do more to translate biological discoveries into human health.

One day, while scaling up the synthesis of RNA, in addition to the expected results, a new molecule appeared, and she kept it for further analysis. Then—and she recalls the exact moment and street along which she was walking—she had a sudden flash of insight. It simply dawned on her that what was in her test tube was nucleoside chlorophosphite, which turned out to be an exquisite starting material for the rapid syntheses of many phosphorus derivatives, including phosphoramidites and thiophosphates, which have been used extensively in the chemical synthesis of DNA and in drug development. Mona also pioneered the use of silica gel as a solid phase. She developed support for RNA and DNA syntheses, and this approach, combined with the use of phosphoramidites, enabled the development of the "gene machine," the automated technology for synthesizing DNA and RNA that makes genetic discoveries that would have taken years possible in very short amounts of time.

Mona's accomplishments as a graduate student (1977–81) opened several opportunities. After spending a year in a US biotech company, she decided to accept a postdoctoral fellowship in molecular biology at the Institut de recherches cliniques de Montréal (IRCM) and Columbia University. During this time, she acquired expertise in gene isolation and gene expression and regulation, which enabled her to make important discoveries throughout her career, in which she consistently demonstrated a propensity for innovation and her desire to explore paths that others had dismissed. For example, during this time she was the first to isolate the human gene for a recently discovered hormone that regulates the balance between salt and water in the body—atrial natriuretic peptide (ANP)—and to demonstrate that it was produced in the heart. Her work contributed to a paradigm shift by establishing that the heart is not simply a pump, but an endocrine organ, actively involved in regulating blood pressure and electrolyte balance. Her discovery in the late 1980s, that within the heart, levels of ANP and related BNP, the hormone indicating enlargement of the heart, increase during cardiac stress and in cardiac hypertrophy, which is the growth and increase of muscle cells, formed the basis for the development of diagnostic blood tests for measuring ANP/BNP levels to monitor cardiac stress in cases of heart failure and after chemotherapy.

These experiences led Mona to the realization that discoveries lie at the intersection of different disciplines, such as chemistry and biology. She has made interdisciplinarity a part of her lab's culture, where she studies the molecular mechanisms that regulate the genetic program of cardiac cells and the underlying cardiac diseases. Her research has led to the identification of the genetic basis of cardiac birth defects, which occur in 1 to 2 percent of the population. In turn, her findings were translated into diagnostic tests for the early detection of degenerative heart disease. Among other discoveries, she isolated the gene GATA4 for a key diac protein, which is important because it controls the regionalization of tissue immunity. She found that mutations in this gene lead to multiple heart defects. She accomplished this while most other researchers were focused on different pathways, assuming similarities between skeletal and cardiac muscles and stress. Instead, her work established closer and unsuspected similarities between heart and blood cell regulation. Innovation is definitely the path less travelled, and this is one of the insights and discoveries of which Mona is most proud. Rather than concentrate on the heart as a muscle, she thought of the heart as a different organ, one that produced endocrines. Her work established closer and unsuspected similarities between heart and blood cell regulation. GATA4 turned out to be essential for converting stem cells into beating cardiac cells and for the survival of adult cardiac cells. These discoveries are being used in R&D of cardiac therapies.

Having worked as professor of pharmacology at the Université de Montréal and as director of the Cardiac Development Research Unit at the Institut de recherches cliniques de Montréal (IRCM), Mona was recruited in 2006 by the University of Ottawa, where she was named professor of biochemisty in the Faculty of Medicine, director of the Molecular Genetics and Cardiac Regeneration Lab, and vice president of research. She stresses the importance of good communication across disciplines and the necessity of creating a milieu in which all have the freedom to pursue their ideas and to be innovative. Her two terms as vice president of research were characterized by expansion in research infrastructure, including university-wide core facilities that enable sharing and innovation as well as multiple centers and institutes bridging technology and social issues.

When the position of chief science advisor to the prime minister was announced in 2017, Mona was attracted by the challenge not only of bringing disciplines together but also of creating a dialogue among academic and government researchers, of ensuring that science is included in policy decisions and that government science is known and open to the people of the country, of making Canadian science known around the world, and of bringing international partnerships and opportunities to our country. In 2017, when she was selected for the position, she became conscious of what it represented for others: for women, immigrants, researchers and their students, as well as for the public. She is aware that she is forging a path for them.

During her mandate, which has been twice renewed (in 2020 and 2022), she has led the creation of the *Model Policy on Scientific Integrity* (2018), the *Roadmap for Open Science* (2020), and the COVID-19 Expert Panel (2020), which provided valuable and timely advice to the government during the pandemic.

When asked what advice she would offer others, she says that people in general are not always conscious of their capacities and women tend to underestimate themselves. If a woman reads a job description and has less than 100 percent of the listed qualifications, she is unlikely to apply. But nobody is perfect, and everyone should accept challenges and acknowledge the fact that they will learn by taking on difficult tasks. Mona reminds us that entrepreneurs are expected to fail, pick themselves up, and, employing all they have learned, try again.

Mona plans to continue to bring together researchers working in different fields and in different milieux, from government to industry, universities, hospitals, polytechnics, colleges, and CEGEPs, in constructive dialogue. Canada is building a strong policy of support for science and innovation. We need to celebrate the brilliance of the members of our scientific community and consider them a precious resource. They represent, like our natural resources, the wealth of the nation. Mona says that Canada's brains (*les cellules grises*), not only our forests, are important to nurture and to recognize. In sharing her innovative vision for the future, Mona offers us a fine example of a dedicated researcher, administrator, and national leader who employs her talents to serve humanity, her colleagues, and her country.

11

ALAN BERNSTEIN

Thinking Big and Innovating

Toronto, ON

An innovative researcher in the field of genetics, Alan Bernstein is renowned for his groundbreaking research on cancer, for transforming health research in Canada, and for the global impact of his work. A fellow of the Royal Society of Canada, he received many awards, including the Royal Society's McLaughlin Medal, and he figures in the Canadian Medical Hall of Fame.

Alan was born in 1947 in Toronto, where his parents settled after emigrating from Eastern Europe. While his parents were not able to benefit from extended formal schooling, his father was well read and could have been a history professor, and his mother was very smart and had high expectations for her children. They considered education most important. Alan did well in school, skipping two grades, so that by the age of sixteen he was already in his last year of high school. During that time, he had won the gold medal in math and physics and liked science, but he also enjoyed music and played the cello. His music teacher took him to a concert by the brilliant Hungarian American cellist János Starker, hoping that the performance would inspire Alan to want to emulate Starker. Instead, it had the opposite effect. Alan wanted to be the best in his field, and he thought he might never be as good as Starker. While he decided not to pursue a career in music, he did continue playing with the student orchestra until he graduated from high school. To this day, Alan maintains that eight years of piano lessons and several years of cello were very good for developing his young brain.

In 1963, Alan went to the University of Toronto, where he enrolled in the math, physics, and chemistry program. After taking six math courses by the end of his second year, he decided not to continue with math. He remembers walking by Harold Johns's office and knocking on the door. Johns was head of the physics division of the Ontario Cancer Institute, and was head of and professor in the Department of Medical Biophysics at the University of Toronto. Usually working in his lab at Princess Margaret Hospital, he happened to be in his university office and Alan ended up talking to him, a conversation that resulted in a summer job in his lab. After asking Harold what exactly medical biophysics was, Alan still recalls his answer: "It doesn't mean a darn thing, so we can do whatever we want. Just avoid painting yourself into a corner." Alan subsequently changed his major to medical biophysics.

In 1968 at the University of Toronto, Alan began his PhD under Canadian biophysicist Jim Till, who co-discovered stem cells with Canadian cellular biologist Ernest McCulloch. Alan was intrigued by the mathematics of genetics, and Jim gave him free rein to work on his thesis, while sharing valuable lessons in academic administration. In 1972, Alan completed his thesis and had already published several papers in academic journals. When it was time to apply for a postdoc, Jim asked him where he preferred to live: London or Paris. Both had good research teams, with London specializing in cancer and Paris in neuroscience. Alan chose London, and in 1972, he went to the Imperial Cancer Research Fund, where he was part of a community of scientists who were at the threshold of understanding basic genetics. In those days, the technology to locate and process genes did not exist, so they used chicken tumor viruses. Chickens have fewer genes. Alan says, "By understanding simple organisms, we thought we would be able to figure out the human pathology." Their work completely changed cancer research.

On returning to Toronto in 1974, he worked at the Ontario Cancer Institute and the University of Toronto. From 1994 to 2000, he was head of molecular and developmental biology and director of research at the

Samuel Lunenfeld Research Institute at Mount Sinai, and he collaborated with Dr. Tak Mak, the world-renowned Canadian geneticist who worked on cloning the viruses that caused bulimia, an eating disorder. They were early pioneers of cloning in cancer research. This innovation led to further exciting discoveries. Alan read all the bioscience articles in the journals, and one day he came across a paper that inspired him with an idea of how to clone a gene. He returned to his lab and told his postdocs, "I think I know how to do this." And indeed, he did. Cloning is fundamental to the development of treatments for diseases like cancer and diabetes and for the eventual ability to repair injured cells in spinal cords, for example.

Alan was invited in 2007 to be the executive director of the Global HIV Vaccine Enterprise. He spent four years with stints in Africa and China and was invited to speak around the world. This experience made him see Canada in a different light. He noted that different cultures have different ways of thinking about science.

Henry Friesen, the outgoing president of the Canadian Institutes of Health Research (CIHR), a national organization headquartered in Ottawa that funds medical research across Canada, had articulated a holistic view of research and a new, virtual institute structure. When Alan was selected to follow Henry as president of the organization in 2012, his job was to polish, improve, and implement the plan for the new organizational structure. When he began, there was a debate among the disciplines, with those in the humanities and social sciences feeling left out of the new structure. He listened to the community and showed that they had been heard by appointing half the directors from the social sciences and including a new Institute of Aboriginal Peoples' Health. He considered science policy an important issue and had learned that program policy and government policy have different drivers, both of which must be understood. His innovations were to embrace diversity and to include all disciplines.

In 2012, Alan's next challenge was to preside over the Canadian Institute for Advanced Research (CIFAR). His vision for CIFAR has been to identify talented people and give them the resources they need to "get on with it." CIFAR has two principles: "Think Big" and "Recognize Excellence." The organization has lived up to its principles, achieving remarkable results in artificial intelligence and quantum computing, for example. Alan notes that the star professors in these fields should be nurtured and they will attract others who want to work with them. The provinces and the universities need to invest in branding and be more innovative and entrepreneurial so young people will be inspired to remain in Canada. Alan encourages the moon shot option over incremental science.

"There is nothing more boring than looking at a decimal point," he says. "The new technologies and breakthroughs are most exciting." He cites the fast radio bursts discovered by the CHIME group, along with CRISPR-Cas9 genome-editing technology, AI, and quantum, and concludes, "We have many reasons to be proud."

Alan considers Global Scholars his most innovative program, and the one of which he is most proud. Dedicated to helping young scholars who are ready to question the status quo and explore new ideas and new ways of doing things, the Global Scholars program includes mentorship by alumni, participation in a network, and interaction with other Global Scholars.

Alan's advice is "Do not follow a straight line. Experiment with what you really want to do and think big." He is amazed at the level of excellence and the innovation of young scholars. His own career has not followed a straight line, going from genetic research to organizing support for health research, to promoting big ideas and young scholars. Currently the distinguished visiting professor and director of Global Health at the University of Oxford, he is a spirited innovator and a visionary leader who supports big ideas and the truly excellent scholars who conceive of them.

12

TERRY-LYNN YOUNG

A Genius among Geneticists

St. John's, NL

If she had not been born in Penetanguishene, Ontario, Terry-Lynn Young would be one hundred percent an Islander. After all, she grew up in Twillingate, Gander, Corner Brook, St. Anthony, and Lewisporte, Newfoundland, and she never pictured herself living anywhere else. There was, however, one thing that could account for the three years she spent in Seattle. That was her passion for science and, in particular, the science of genetics.

As a child in the sixties, she remembers that her mother's career—as a nurse, then a director of nursing, and later a nursing home administrator—uprooted the family as job opportunities presented themselves. Her father, a watchmaker and jeweler, opened shop wherever they settled. Terry-Lynn's love of biology came about in part thanks to her high school biology teacher, Hector Pearce, who used to leave the lab open so students could do experiments at night. Throughout her childhood, she had always collected "bugs and slugs," and her love of nature went so far that she took the goldfish with her when her family went on holiday.

She went on to complete her undergraduate degree in biology in 1985, a master's in human genetics in 1987, and her doctorate in the same field in 2000, all at Memorial University in St. John's. During the summers, she worked at the Vector Pathology laboratory picking nematodes, which are a type of roundworm, and placing them on slides. Later, during the course of her studies, she was working as a research assistant to Roger Green, a professor of human genetics, when they located a chromosome for colorectal cancer. That moment had a huge impact on Terry-Lynn. She remembers being in the lab with Roger and unrolling the huge pedigrees (family trees) across the floor, identifying individuals with the chromosome that put them at high risk of developing colorectal cancer. They penciled in the genetic marker data right on the paper copy. Her elation at the discovery was overwhelmed by a sense of urgency, as she knew they had to contact the people they identified who were at high risk as soon as possible. It was at this moment that Terry-Lynn knew this was what she wanted to do with her life—genetic research, which was, and remains, her passion.

Following the completion of her PhD, Terry-Lynn took a course in medical and mammalian genetics at the JAX labs in Bar Harbor, Maine. On the way back home, she met with renowned American geneticist Mary-Claire King, a breast and ovarian cancer researcher and one of the first to posit that cancer might be inherited. In 2000, Terry-Lynn was offered a post-doctoral fellowship to study with Mary-Claire in her lab in the Department of Genome Sciences and Medical Genetics in the Department of Medicine at the University of Washington in Seattle. Terry-Lynn worked there for three years and was then offered a position at Memorial University, which she refused to accept until the university agreed to provide her funds for lab equipment. She could hardly return to Newfoundland to work on genetics without a sequencer.

In 2003, she returned to Newfoundland, where she was made aware that a very high percentage of young men were dying suddenly of cardiac arrest. Working with Kathy Hodgkinson, a professor of genetic medicine, and Sean Connors, a professor of cardiology and epidemiology, her team identified a lethal gene mutation, TMEM43. They started creating a genomic map of individuals who had passed away from sudden cardiac arrest. Hodgkinson had already been putting together family trees. They soon realized that they could identify the individuals likely to have this genetic mutation. Half of them would probably die before the age of forty. With her colleagues, Terry-Lynn identified carriers of the lethal genetic mutation, helped develop screening methods, and found preventative treatment: a life-saving implantable cardioverter defibrillator. They crisscrossed Newfoundland and Labrador to locate family ancestors in graveyards, connecting families that were related to each other and offering those with the genetic mutation to have the device implanted. They saved the lives of hundreds of carriers of this gene. Terry-Lynn recalls one Islander with an implant who was at a cabin with his buddies. Suddenly, he felt a jolt and fell from the sofa to the floor. Thanks to the device, he survived the episode of arrhythmic heartbeat and his pals had quite the story to tell.

When she was a student in St. John's, she noted people in the street with signs indicating that they were deaf and mute, seeking spare change. They appeared otherwise hale and hearty, but could not find jobs due to their condition. Her PhD was on the genes that caused kidney disease and blindness. She now wondered if hearing loss was also genetic and found the gene that causes otosclerosis, a common form of conductive hearing loss, which is often hereditary. In 2016, she went to Grand Falls–Windsor, located five hours from St. John's, to meet with a local audiologist, Anne Griffin. Terry-Lynn then set up a centre where hearing loss could be treated. People would often arrive in tears, but leave the clinic feeling happier. Training volunteer high school students, who became interested in a career in audiology, was an additional innovation.

In recent times researchers have focused on sequencing the whole genome and its variants in an individual, thereby losing the powerful insights that could additionally be gained through linkage analysis, tracking variants through extended families. In Newfoundland, large families were common, and Terry-Lynn felt they should look at them for patterns that might otherwise be missed.

"Everyone has thousands of [genetic] variants," she explains. "It is difficult to figure them out individually, but when you track them across time in large, affected families, you can find genes you would never find otherwise."

She notes that today, families are growing smaller, so there is an urgency to identify genetic disorders or diseases now while the control groups exist.

Establishing trust is important. In Newfoundland, the community leaders are the doctors and the representatives of the church. Terry-Lynn credits her predecessor Jane Green, professor of medical genetics at Memorial University, and her colleague, Kathy Hodgkinson, for their work with families. To their insights, Terry-Lynn added the technology and lab needed to support discovery research. She met with families in fire halls and stayed in town with the team audiologist, Annie Griffin. She went to the homes of people who needed appointments. Outside the hospital, Terry-Lynn and her staff wore jeans and T-shirts in order not to intimidate people.

Terry-Lynn's next initiative is to work on balance issues, which she has observed in some families in the region, especially in the elderly. She wonders if there is a connection between balance and hearing loss. She is also partnering with the Miawpukek Mi'kamawey Mawi'omi First Nations near Grand Falls–Windsor to help the community connect to research projects on hearing loss and heart disease. Through their collaboration, the community will own the data sets. In addition, she is working with a national team to identify the role of hearing-loss genes in participants in the Canadian Longitudinal Study on Aging (CLSA).

Her advice to young scholars is to be persistent, to "figure out what you want to do. You do not need to go to the biggest university. Some issues are best studied on the community level." She recommends an interdisciplinary approach, and finds, for example, that psychology and sociology play an important role in all scientific research. She recalls a team of researchers from Texas who arrived in Newfoundland by helicopter to study sudden cardiac deaths. They took blood samples, left, and never reported back. This did not build confidence or trust.

Terry-Lynn loves her work and hopes that one day there will be a major genomics centre in Newfoundland. She sees applications in her field that will become increasingly important in other areas such as mental health and other diseases where there is likely a genetic connection. The prospects are exciting. She feels compelled to keep going and knows there are always problems waiting to be solved.

13

GARNETTE ROY SUTHERLAND

Revolutionizing Neurosurgery

Calgary, AB

Garnette Sutherland, a neurosurgeon, has received many awards for his trailblazing work in medical technologies. Named one of the top ten achievers by the Canadian Institutes of Health Research (CIHR), he was recognized by the American Astronautical Society for earth applications of space technology and is in the Space Technology Hall of Fame. His most famous innovations to date are the world's first intraoperative magnetic resonance imaging (MRI) system based on a ceiling-mounted movable high-field magnet, and another world's first, the neuroArm, which is an image-guided MRI-compatible robotic system for brain surgery.

Born to Audrey, an accountant who grew up on a farm north of London, Ontario, and Bert, a farmer from Saskatchewan who served in the Canadian Armed Forces, Garnette had a peripatetic upbringing, following his father's various postings in France, Nova Scotia, and Manitoba.

Garnette had always been interested in biology and science and was fortunate to have a very gifted advanced calculus teacher. He says she was "rather eccentric, but could translate the beauty of calculus to her students," and for Garnette, this feeling extended to the love of all science. Between semesters, he would fly to Toronto and apply for a job with an insurance company. He always found a job within a week and would return home in time for the start of classes in the fall.

An industrious student, Garnette earned both his BSc and MD degrees from the University of Manitoba, Winnipeg, in 1974 and 1978. As an undergraduate he worked in the lab of physical chemist Ernie Bock, who encouraged him to pursue his studies. While in medicine, he did his third-year elective rotation at the Montréal Neurological Institute, where he met Theodore Rasmussen, the famous neurosurgeon who specialized in epilepsy, and after whom the Rasmussen syndrome (a chronic form of inflammation of the brain that often accompanies epilepsy, a brain disorder causing seizures) is named. Garnette recalls also being inspired by Dwight Parkinson, Canadian neuroscience pioneer at the University of Manitoba. On graduation, Garnette completed his neurosurgery residency at the University of Western Ontario (now called Western University) in London, Ontario, in 1984. There, he recalls the daunting experience of treating patients with complex neurological diseases, using the best technology available and dealing with the possibilities of complications. He remembers Charles Drake, the legendary neurosurgeon known for his work in treating aneurysms. He was at the time also the head of the division of neurosurgery and would often say, "If I could do this case again, what would I do differently?" Garnette particularly admired Drake's exceptionally skilled touch as a surgeon, which inspired him to invent a way to measure and teach this skill.

His early work on the effect of trauma on the heart, and its systemic sequel using novel vascular labelling, led to high-impact publications and interest in the application of similar techniques to neurovascular conditions. Thus began his academic career, forging innovative solutions to complex diseases.

On completing his residency, Garnette was recruited by the University of Manitoba as an academic neurosurgeon, with a co-appointment and laboratory in the Department of Pharmacology and Therapeutics. His collaboration with other scientists interrogating models of neurological disease resulted in the relocation of the National Research Council's (NRC's) imaging division to Winnipeg, Manitoba.

Subsequently, upon relocating to the University of Calgary as the chair of neurosurgery, Garnette began the process of adapting magnetic resonance imaging technology for the operating room. Shaped by his observations and work with Charles Drake, Garnette went on to propose his first major innovation: the previously mentioned magnetic resonance imaging system that operated on a robotic arm. This technology, an upside-down magnet on wheels, seemingly inconceivable at the time, was realized and is now located in more than seventy-five international sites, benefitting over seventy thousand patients, thanks to global collaborations and commercialization.

Garnette continued to consider the challenges and the nuances of intraoperative MRI. At the time, neurosurgeons had to stop surgery to acquire images. This disruption during intense brain surgeries was the impetus for the invention of neuroArm, the MR-compatible telesurgical robot, capable of microsurgery and stereotaxy, which is a technique that records and reproduces three-dimensional touch information to create the impression of depth. While searching for the ideal partner to bring this transformative idea to fruition, Garnette had several conversations with one of his residents, Taro Kaibara, which led to a connection with MacDonald, Dettwiler & Associates (MDA) in Brampton, Ontario. With the launch of Canadarm, Canadarm2, and Dextre, and MDA's impressive achievements in space robotics, Garnette thought, *If they could build a robot for the hostile environment of space, surely they could build one for the neurosurgical operating room.* This new robot would have to work in the restricted three-dimensional space of a magnet bore and be able to function during imaging, which would be commanded by the surgeon at a remote workstation, where the sight, sound, and touch of surgery could be re-created. Garnette was venturing into unexplored territory. He was bringing aerospace ideas into medicine, and with MDA as his partner, he once again "garnered extraordinary support from the Calgary philanthropic community, Western Economic Diversification Canada (now Prairies Economic Development Canada), and the CFI, known to fund the Big Ideas that change the scientific landscape in Canada." The result was the world's first MRI-compatible robot, a one-off system well ahead of its time that would serve generations of engineers, entrepreneurs, and surgeons around the world.

Garnette continued his innovative thinking. He would often recall Charles Drake and the technical finesse with which he could dissect an extremely small lesion in the brain without tearing or breaking it. Peers and trainees would watch him work, but could not see the amount of pressure he actually applied with his surgical instruments to repair fragile lesions and structures. Just as Garnette's neuroArm enabled the very precise measurement of force in Newtons (the product of mass and acceleration),

Garnette challenged his engineering team to imbed sensors in common surgical tools, in particular the bipolar forceps, to measure and report the amount of pressure the surgeons applied as they operated. Through data science and digital innovation, this became a system, called the SmartForceps System. With built-in machine learning and cloud computing, the SmartForceps System provides surgeons with a performance report. They can see the amount of pressure they applied and the results. This means that surgeons can look at their performance and note ways to improve it, and residents can see the amount of pressure that was applied in different situations. No similar technology exists elsewhere, and Garnette believes that this innovation will improve the art and technique of surgery, while accelerating learning.

Garnette is working on the next generation, called CellARM. He would like to increase precision at the cellular level so that only abnormal cells would be removed. His work with the National Research Council and the University of Victoria on delivering antibodies to abnormal cells with fluorescent markers, and with university collaborators across the country on cell vibration, will enable surgeons to "see what they cannot now see and hear what they currently cannot hear." His commendable work on haptics, or touch, continues to move forward. Unique software interface and machine learning could transform the neuroArm to a robot that approaches or parallels the executive capabilities of the surgeon in the loop. Garnette calls this invention Cloud^{ASSIST}.

Garnette says that, for him, innovation is more of an evolution of ideas over time that involves integrating collective knowledge and experience in solving a daunting problem. It is generally a slow process. There is a period of information gathering, followed by trial and error in creating a model that works. As a practicing surgeon and team leader of the large research space Project neuroArm, he touts the merit of the lab meetings, hallway run-ins, and discussions of issues over coffee that bring people together to present ideas and solutions. When a group gathers, the members may change part of a concept or do something a bit different, and suddenly a new pathway opens.

Garnette cautions that the process does not end when the solution is in sight. A machine has to be built and it must pass all safety regulations. He recalls having built three prototypes and having been able, only after the third, to produce a commercial run. Garnette firmly believes that innovations in labs can and should become viable products, so that not only patients, but the Canadian innovation and economic ecosystems benefit as well.

With an endless supply of challenges and possibilities on the horizon, Garnette will not lack interesting problems to solve and a universe full of innovative solutions to discover.

14

JAMES ROBAR

Inventing Better Care and Treatment

Halifax, NS

James Robar learned a thing or two about innovation at his father's knee. His dad was a physicist, specializing in atmospheric and space science, who worked in Saskatoon for a company that built and sold satellites. When the company was no longer profitable, it turned to making sensors and machines for milking cows—another example of innovation at work.

When James was ten, his family moved to Ottawa, Ontario. When he was not at school, he attended Cub Scouts. With his father, he built the traditional cub car. The idea was to carve a block of wood into the most aerodynamic shape possible and then win a race. James decided not to compete for speed, but to go for the style category. He and his dad hollowed out the bottom and installed flashing lights. Their cub car was quite spectacular. Unfortunately, in removing much of the bottom, they accidentally cut the axles. As a result, the car barely moved three feet and sat on the track twitching and flashing its fancy lights.

James says he was most fortunate to attend an excellent high school, Lisgar Collegiate, in Ottawa, from 1985 to 1990. He fondly remembers participating in science fairs. One of his projects was to calculate the thickness of the wall of a soap bubble. He made his calculations by comparing the bubble to a skydiver. When you know the terminal velocity—its speed on hitting the ground—you can create an equation to figure out the actual mass of the object. James said the only downside to this experiment was that he was "sticky" for months.

He also recalls the time in physics class when he built a cloud chamber to visualize alpha or beta particles. He remembers soaking a piece of black velvet with alcohol and putting it under a glass mixing bowl he borrowed from his parents' kitchen. He then turned the bowl upside down over the velvet before sealing the edges, resulting "instantly in a massive fire with flames shooting five feet in the air." Luckily, his teacher had a cool head and extinguished the flames, and James was happily not deterred from conducting more experiments.

James had quite a few summer jobs over the course of his studies. One of his favourites was at a downtown hardware store in Ottawa. People would come in and tell him about their home project and describe their problem. James would listen and make recommendations on what tools they needed and explain how to fix the problem. This service resembles what some engineering faculties offer today as a way to build community relations and give students practical experience.

He also worked at a bakery, where he started at the counter and eventually made his way up to manager. He hired all his friends, and together they made fabulous sandwiches—another example of the precocious young innovator at work.

James's parents were always supportive and did not pressure him to attend university. Their mantra was simply "Do what you love." But James says that with two such brilliant parents—his mother also worked as a pharmacist—they led by example.

In 1990, he arrived at McGill University in Montréal, where he had been awarded a Natural Sciences and Engineering Research Council (NSERC) scholarship, which enabled him to work in the summers with Dr. C. J. Thompson, a professor of medical physics, who was building a particle detector as part of a quark-gluon plasma project at Brookhaven National Lab in Upton, New York. While their efforts represented only a tiny part of a huge experiment, James was invited to spend two summers at the facility in 1992 and 1993. It was a rarified atmosphere. Only physicists and their families were allowed to enter the high-security facility. James's job was to work nights, "babysitting" the multiwire proportional particle counter. If it stopped, James was supposed to figure out how to fix it.

While at Brookhaven, James missed the interdisciplinarity of the university and the academic community where he was studying both physiology and physics, but he was not sure what to do after graduation. He was definitely interested in medicine; he did not, however, enjoy rote memorization. On the other hand, physics was "just trying to understand how the world works," which James rather liked. In the midst of his uncertainty, there was a terrible snowstorm in Montréal. That evening, the buses were out of service, and he had to walk home. It was a long hike, and his route took him past a building with a sign that read "Medical Physics Building." He later returned and visited the faculty. One of the professors, Dr. John Schreiner, chief of medical physics (now at Queen's, but then at McGill), invited him to lunch. Over smoked meat sandwiches, James learned that if he studied medical physics, he could be a pure researcher, a clinician, or work for government or industry. The field includes physicists, engineers, and radiologists, who use instruments and techniques including ultrasound, X-rays, nuclear medicine, and radiation therapy for diagnosis and treatment of diseases such as cancer. James found Schreiner inspiring, and he decided that he had accidentally come upon the perfect place to study, especially when combined with the offerings of the Montreal Neurological Institute. On completing his MSc in 1997, James went on to do his PhD at the University of British Columbia (UBC), Vancouver, in medical physics, completing his medical degree in 2000 and his residency at the BC Cancer Agency in 2003. Professor Schreiner still remains his friend and collaborator.

James's work has resulted in an extremely impressive series of innovations. For example, years ago, when women were screened for breast cancer, a biopsy was very frequently recommended. The great majority of biopsies were negative, and patients had been subjected to an unpleasant process. James's former professor, C. J. Thompson, had an idea about the PET scans that use radioactive substances to visualize and measure change in the body, and James set out to construct an imaging system that could be used for mammograms. In the end, his research team could detect both tumors, distinguishing those that were benign from those that were malignant. In the space of a year, they accomplished this breakthrough, and the technology was adopted, saving women discomfort and concern, as well as their lives.

While at UBC, he moved from medical imaging to radiation oncology physics for treating cancer. With his supervisor, Dr. Brenda Clark, chief physicist at the BC Cancer Agency and professor of physics at the University of British Columbia, he worked on radiosurgery for brain tumors, using high-energy X-ray beams that converge from different angles, overlap on the tumor, and kill cancer cells, while harming as few healthy cells as possible. Their innovation was to start shaping the radiation beams to match the exact shape of the tumor in the brain. They also wanted to reduce toxicity. James had to figure out if submillimetre doses of radiation would be safe and efficacious. A special tool would need to be created, so he invented what his wife called "Aquaman"—a transparent three-dimensional replica of the human head that was filled with water and had a 3D detector in the middle of it. By using "Aquaman," they could determine where the dose went and make measurements, which were then fed into the computer prior to surgery.

James says that UBC was well set up for tech transfer and taught students how to write patents, and they were encouraged to obtain licenses and start companies or spin-offs. This is how "Aquaman" was licensed, and it was James's first exposure to applied science. He had always felt that it was not enough to write a paper and "hope for the best." He feels the impetus to translate his research into practical application.

In 2003, James and his wife began thinking about settling down. They loved sailing and wanted to live by the ocean, but the housing prices in Vancouver were prohibitive. During this time, James was invited to give a presentation in Halifax, Nova Scotia, and, with his wife, decided to move there, where he could walk to work and they could still enjoy sailing.

James was then invited to join Dalhousie University. He established the master's, PhD, and certificate programs, which were approved by the Dalhousie senate in 2014 and received official accreditation in 2015. They serve as an engine for R&D. While he is not paid by the university to teach, James felt a responsibility to train the next generation in the creative process, as no such programs were available east of Québec. The programs have since grown by a factor of five.

After building an environment that would foster innovation, he next turned his attention to 3D printing. Around 2012, the original patents for the printers had expired, so everyone could build them, making them less expensive and more accessible. James thought of a possible application to radiation therapy. Until then, patients requiring radiation therapy had to lie still while a mold, called a bolus, composed of pink paraffin was made directly on the surface of their body. Sometimes there would be an air bubble or gap beside the patient's skin, and the mold would have to be recast. Being covered in hot wax was not a pleasant experience for patients. In 2013, James realized that he could use computer data to generate an accurate 3D mold.

James had a fantastic graduate student, Shiqin Su, who did the algorithms for the device to tailor the radiation inside patients, as well as create the mold. James and two business partners, including current CEO Alex Dunphy, created a spin-off company called Adaptiiv Medical Technologies. The bolus idea gave rise to an array of different technologies in the field, and Adaptiiv's products are used in fifteen countries around the world.

Yet another device resulted from what James says was an accidental discovery. In treating a brain tumor, it is crucial that patients do not move, and up to that point, patients wore a mask to remain immobile. Doctors needed real-time detection and an instrument that could see through the mask. James found a conductive paint that was used for "painting" circuits. Finding the paint may have been a coincidence, but seeing what he could do with it was a brilliant innovation. He painted an array of paths in a ring, connected the detectors, and wrote code to form the output into a display. He asked his five-year-old daughter to put her hand in the ring and draw. A screen then displayed where she was moving her finger. This was the perfect solution—doctors could see in real time the precise movement of the body projected on the screen. This technology has been licensed to an industry leader in the radiosurgery space.

James says that his research and innovation are ways of caring for patients. Research and clinical practice go hand in hand, and both play essential roles. He has many dreams about new technological advances and, although he does not believe we will eliminate cancer, he thinks we will be able to treat it more effectively. He dreams of the day when all cure rates are 80 to 90 percent and the fear of cancer is gone.

James's innovations have saved lives and diminished suffering, and he will, to be sure, continue his trajectory of success. The next time there is a big snowstorm, and the buses are not running, I, for one, will think of James, and will be grateful for the inclement weather that led him to the Medical Physics Building.

15

SHEILA SINGH

Engineering Cells to Cure Cancer

Hamilton, ON

Born in Hamilton, Ontario, Sheila is a first-generation Canadian. Her mother was from Austria and her father from India, and they met in England, where her mother was studying nursing and her father was training as an MD in psychiatry in the 1960s. They were both working in a hospital, and when they finished their training, they were married and set their sights on Canada. Her father got a job in Prince Albert, Saskatchewan, and Sheila's mother, accustomed to the Alps in Austria and the rolling hills in England, was depressed by the immense and unending flatness of the landscape, which is how they ended up in Hamilton, Ontario, in 1971.

From the age of two and a half to eighteen, Sheila attended Hillfield Strathallan College, which provided a welcoming and warm community where Sheila made many friends. She credits Hillfield, which used the Montessori teaching methods, structure, and rigor combined with the freedom to choose activities, with having sparked her creativity and her ability to work in self-directed circumstances. She learned broadly across disciplines, including music, chemistry, biology, drama, and English. She particularly loved the classics. Years later, when she returned to Hamilton with her two children, she enrolled her sons in the school, where her former teacher is now the head.

In summers she would divide her time between work and travel. At the age of fifteen, she began working in the Ram Mishra neuroscience lab in Hamilton as a volunteer washing test tubes, and was later assigned more complex tasks. Her travels took her to England, Austria, and France, where she worked as an au pair and learned French. She was rather precocious, reading Freud's *The Interpretation of Dreams* at the age of ten. Another memory from her youth was of the Nike slogan, "Just Do It," encapsulating her activist nature, which had included protesting against apartheid outside the Broadway Theatre in Hamilton at some point during Nelson Mandela's twenty-seven-year imprisonment in South Africa.

She studied neurobiology and genetics at McGill University, where she volunteered at the Montreal Neurological Institute-Hospital, which piqued her interest in molecular genetics and neuroscience. The brain seemed like a vast, unknown territory. By her third year, she was allowed to view open craniotomies and other brain surgeries from the gallery of the operating theatre. She recalls the "beautiful surgery with elegant procedures used to map the brain to find the focus of seizures." Following her undergraduate degree, she went back to Hamilton to attend medical school at McMaster University. Graduating in 1997, she completed the program in two years. She took every course offered in neuroscience and learned how to map the brain with an MRI and find lesions.

She was immediately accepted at the University of Toronto for residency training, and her fiancé went to Waterloo University for architecture. After two years, they got married while she was doing her residency at Sunnybrook Hospital in Toronto. She says she truly cares for all her patients and that neurosurgery is amazing because patients can be successfully cured, but there is always a risk of death, and "every surgeon carries in their heart a graveyard they visit from time to time." She shared a moving story of two patients whom she observed in the hospital at the time. Both were five years old. Both were named Christopher and had the same tumor, a medulloblastoma. Both had chemotherapy, radiotherapy, and surgery. One survived and did well and the other died. Nobody could understand why. Sheila decided that she needed to study the molecular biology of brain tumors and paused her residency to work and study for a PhD with Dr. Peter Dirks, a neurosurgeon and molecular geneticist at the University of Toronto with a lab at the SickKids hospital. She wanted to understand how brain tumors in children occur—how a normal stem cell in the brain might be transformed into a cancer stem cell.

Dr. Dirks and Dr. John Dick, a molecular geneticist who sat on her PhD committee and had recently discovered leukemic stem cells, challenged her to find out if stem cells in brain cancer tumors could grow in regular culture conditions, without serum. Sheila read about a recent discovery by another Canadian, Dr. Sam Weiss, a prominent neurobiologist who discovered neural stem cells in the brains of adult mammals, which meant that they could generate new cells. He had also found that normal neural stem cells grew in serum-free conditions, and Sheila decided to extrapolate these conditions to brain tumors. She removed the serum and, indeed, the brain cancer cells grew, declaring their existence by the presence of "beautiful spherical colonies" that she found floating in the culture dish. These tumors had not previously been grown outside the body and could now be closely studied. After completing her neurosurgery residency, and a PhD in 2005, Sheila did a year as senior neurosurgery resident and another with a fellowship at SickKids. When she returned to McMaster in 2007 she brought the tumor sphere model system, which is considered one of the best models for brain cell tumors in the world, along with gold-standard stem cell animal models. She says she could not have set up her new lab and clinical practice in pediatric neurosurgery in 2007 without the help of her family. Both her mother and her husband's mother assisted with the children. She also recognizes her trainees and the fine interdisciplinary collaborations that are possible at McMaster.

Sheila says that it takes five to six years to do the "heavy lifting" to set up a lab and develop models before you can actually start meeting specialists in other fields, such as medicinal chemistry, functional genomics, and proteomics. She has been working with CAR T cells, which are T cells (ordinary white blood cells found in the body that fight infection and disease) that are removed from the body and reengineered with CARs (synthetic molecules that target cancer cells) and cause the patient to develop new immunotherapies when they are reinserted in the body. Sheila engineered a new CAR T cell that reacts against a glycoprotein that is found in certain types of brain tumors and whose presence is an indicator of cancer and its profusion, a sign of the progression of the disease. This transmembrane glycoprotein is employed as a brain cancer stem cell marker, CD133, and is currently being developed commercially, and Sheila hopes it will one day save lives.

Sheila is an academic scientist at heart and loves passing on her knowledge to students. She is also passionate about research and discovery, and likes being "part of the pipeline" and seeing her work help patients. She is co-founder of Empirica Therapeutics, along with functional genomics and CRISPR expert Jason Moffat at the University of Toronto. This company focuses on the rational discovery of CD133-targeting immunotherapies. The company was acquired by Century Therapeutics in June 2020, resulting in the creation of a Canadian subsidiary. Her postdoctoral fellow, Parvez Vora, also a co-inventor, went with the company and now serves as the head of Century Canada Laboratories. She herself has chosen to remain in her lab, collaborating with colleagues at Stanford University in California and St. Jude's in Memphis, Tennessee, on engineering the next generation of CAR T cells.

She equates innovation with creativity that allows us to explore and to develop new insights.

Sheila's advice is to "be bold and brave; you know you are on the cutting edge when you are out of that comfort zone." She also encourages researchers to collaborate and says emphatically, "Today, science is a team sport."

Sheila's modestly "simple story of growing cells in a petri dish" led to the possibility of solving complex problems and moving medical treatment of brain cancer to the next frontier. Her story is also one of a lifetime commitment to science.

16

DANICA STANIMIROVIC

Crossing the Blood-Brain Barrier to Treat Disease

Ottawa, ON

Danica Stanimirovic's dream was to study neuroscience, and she took a somewhat circuitous route to arrive at the path that is leading her to do innovative, groundbreaking work that will save countless lives.

Born in 1962 in a small town in former Yugoslavia (now Serbia), Danica and her family moved to Belgrade to be near the university, as her father wanted her to follow in his footsteps and become a medical doctor. She thus went on to medical school in Belgrade in 1982 and was the first graduate of her class and the youngest medical doctor at that time. She was, however, really fascinated by the brain and how it works. She was interested in both neurology and psychology and thought about becoming a neuropsychiatrist. So she went back to school in 1989 and completed her master's and PhD in neuroscience. At that time, Yugoslavia, which included Bosnia and Herzegovina, Croatia, Macedonia, Serbia, and Slovenia, was in considerable turmoil with the start of a civil war. On her graduation in 1991, Danica was fortunate to receive a postdoctoral fellowship at the National Institutes of Health (NIH) in the United States, and made the big decision to leave her family and cross the ocean.

She spent two years at the NIH's National Institute of Neurological Disorders and Stroke, doing research on strokes and how the brain functions. The war in Yugoslavia continued and, while it had never been Danica's intention to stay in the US, she could not now return home, and she needed to support her family. She could not stay in the US, either, because her visa was expiring. She applied for a position as research associate at the National Research Council (NRC) in Canada, which was starting a new program in neurosciences, and was accepted in 1993. As the only trained neuroscientist at the institute, she considered it a challenge, but a great opportunity to build a strong research program. Thirty years later, she is now director of R&D in the Translational Bioscience Department at the Human Health Therapeutics Research Centre, working in neuroscience, a field in which she proudly states, "Canada is one of the most cited and innovative countries in the world."

Danica's expertise is the blood-brain barrier. The vessels that bring blood to the brain form a very dense network and provide energy to the brain. The brain needs more than 20 percent of the energy we consume. The brain is also delicate and needs a very stable microenvironment. The vessels and capillaries that provide the brain with nutrients also protect it from blood-derived toxins by forming an impenetrable barrier. Danica says that if she could find a way to direct specific medicines to affected areas of the brain, it would eventually be possible to treat specific diseases such as Alzheimer's or even chronic problems of the eye, like glaucoma.

Danica worked to understand the molecular composition of the blood-brain barrier. In fact, she was the first person in the world to use different disciplines to understand the way these barriers work. She understood that they are integral parts of the brain and recognized that one can "coerce" natural blood-brain barrier transporters to bring treatments into the brain. Over the years, she has applied many technologies that have been developed at the NRC, including using very small antibodies to build a "molecular Trojan horse" that can carry therapies across the blood-brain barrier. She was also the first person in the world to do this.

When asked how she made these discoveries and innovations, she says she creates images in her mind and follows a mental process moving from problem to idea. She looks to see if an idea is feasible and then tests it from different points of view. The idea itself comes from less structured thinking and emerges sometimes while she is driving through nice landscapes, for example. (One of her colleagues, on the other hand, gets ideas while playing video games.)

She is most proud of having discovered antibodies that can carry therapies across the blood-brain barrier, because they will have a tangible impact on developing better, more efficacious medicines for brain diseases. Currently, they are undergoing clinical trials. She is also pleased that she facilitated discoveries of drug therapies for several rare and previously untreatable diseases. She notes that after the first discovery there are always innovations that improve on it, saving additional lives.

She hopes that the field of neuroscience will become open to entirely new approaches to tackle very difficult problems. Over the years, she has enjoyed painting, but has always kept it on the "back burner." Recently, she assembled a group that plays with visual art and neuroscience, neuroscience and music. She is convinced "it is very important to keep these elements in our overall environment." Her sessions provide yet another way to collaborate and exchange ideas. They offer a model for building ideas, just as the artist applies paint. They also are a way to train the mind to see differently and to observe things in a new light.

Her advice to young people is to follow your heart and figure out what you are good at doing. You need strong motivation and must be ready to commit not just to one single project but to working in a changing environment that shapes the nature of your work. Collaboration among researchers in different fields is important. Engineers, computer scientists, and specialists in artificial intelligence working together in a multidisciplinary environment with bioscientists are at the crossroads where innovation will occur. At the same time, people who have come from around the world or who have lived abroad have often learned different ways of seeing things. We can learn from them. She says, "It takes effort to transcend ingrained ways of thinking."

17

MOLLY SHOICHET

From Polymers to Medical Treatments and Cell Regeneration

Toronto, ON

Born in Poland, Molly Shoichet's father came to Canada in the early 1930s. Her grandfather died when her father was nineteen, so he had to set his studies of genetics aside and support his mother and sister by running the family business. Molly's mother was born in Toronto, and her parents also passed away when she was young. Both parents encouraged Molly and her two brothers to do their best, and they were Molly's inspiration and role models. Her mother was an ardent feminist and wanted Molly to do whatever she liked as long as she would have a profession.

From grade five to graduation from high school in 1983, Molly went to the Toronto French School, where she was inspired by her chemistry teacher and a math teacher, who told her she was gifted in mathematics. The teachers encouraged all students to "think big" and do great things. Immediately on graduating from high school, she went to MIT to study chemistry. She joined the undergraduate research opportunities program and was encouraged to do research in both a biomedical science lab and an engineering lab. But it was in her advanced chemistry lab that she became fascinated by polymers, large molecules that can be natural, such as DNA and proteins, semisynthetic, such as rayon, or synthetic like polyester. She employed all her electives to take courses on polymer chemistry. She also used her research project to learn more about these molecules and became inspired about the possibilities in this field to "invent the future and do something better than had ever been done before."

In 1986, Molly applied to medical school, but deferred her admission and went instead to the University of Massachusetts at Amherst, which had the best program in polymer science and engineering in North America. Amherst is a public university, quite unlike the private schools Molly had previously attended, and the town was in a pastoral setting, far from Molly's normal urban habitat. She remembers looking out the window at a pasture with cows and asking herself, *How did I end up living here?* Her supervisor had a fantastic research team and gave Molly great freedom with her thesis. Her focus was at the intersection of applied chemistry and applied biology, and she was so passionate about her work that she decided to finish her PhD and not go to medical school.

On graduation in 1991, she asked her supervisory committee for job suggestions, then wrote letters and made many calls. Eventually, the offers started to arrive, and she took a job at a small biotech company in Providence, Rhode Island, working on stem cells and regenerative medicine. They needed someone in polymers. It was a young company with supersmart, highly motivated people. The company encouraged staff to attend conferences and to publish their findings. Molly published six papers and filed ten patents in three years. She looks back on this job as a high-paid postdoctoral fellowship. She realized that published papers were important for academic currency and has over 650 published papers, patents, and abstracts today.

In 1995, she decided to return to Canada, where the biotech sector had grown. Her lab at the University of Toronto combines chemistry, biology, engineering, and medicine. To be innovative, she muses, you must think in different ways and contemplate different ideas and approaches. She values people who advance knowledge to practical applications. This is true innovation. Science is not only discovery, but also application.

Molly says it is easy to start companies, but hard to grow them. She knows this from personal experience, having started four spin-off companies. Success lies in collaboration. Nothing can be accomplished alone. You need new ideas and experiences. She says her years in the US (from age eighteen to twenty-nine) were hugely formative. She likes the proactive American problem-solving mentality, and she brought it with her to Canada. One of the companies she co-founded, AmacaThera, uses injectable hydrogels to deliver medicines and therapies to a targeted location in the body. Good for acute pain, the drugs are formulated to last longer and thus reduce the likelihood of addiction to painkillers.

Half of her lab works on ways to kill cancer cells, while the other half works on regenerating cells. The researchers use hydrogels to grow cancer cells in a three-dimensional environment, which closely resembles that of the human cell. With this system, her team can better understand disease progression and how it can be treated. Twenty years ago, Molly wondered if cells could be guided to grow in three dimensions, publishing the first paper in this field in 2004. Researchers could already guide cells on flat surfaces, but she speculated that if they could be guided in 3D, we could better understand the basic biology. She adopted the principles of physics and used light from lasers to modify materials. She wanted a structure that would guide cells by chemical and not physical signals. If everything in the pathway of light is changed, she further intuited, you could cause change by combining chemistry and lasers. Now that this has actually been achieved, Molly and her students are looking for further applications.

The other half of her highly interdisciplinary lab concentrates on the central nervous system and specifically on regenerating the brain after stroke, the spinal cord after traumatic injury (paralysis), and the retina due to blindness. They are learning how to transplant stem cells so they survive, and how to make environments conducive to their propagation.

"The final frontier of medicine is the nervous system," she says.

Molly hopes that in ten years, knowledge will have sufficiently advanced that they will be able to help people recover from illness and accidents.

The advice that she gives to the students presently in her lab, and the more than 220 researchers she has trained, is that the most successful work comes from collaborating with others. She advises everyone to "say YES to life. You do not always know if what you choose to do will lead somewhere, but it is worth taking a risk. Put yourself in uncomfortable situations and learn from experience." Molly also believes in promoting innovation and engaging the public in research. She co-founded Research2Reality, which uses social media to inform the public of innovations in science, technology, medicine, and social sciences.

When she studied chemistry at MIT, there were few female faculty members and only 23 percent of the students were women. She was just eighteen and yet she created a space in which she could make a difference. She has always been an advocate for women in science and thinks we have made progress thanks in large part to her mother's generation, who were the trailblazers. Molly says that she did not realize that she, too, would be a trailblazer.

Molly has received fifty awards for her work, including NSERC's Herzberg Canada Gold Medal and the Killam Prize in Engineering. She is the only person to be inducted into all three of Canada's national academies: the Canadian Academy of Health Sciences, the Canadian Academy of Engineering, and the Academy of Science of the Royal Society of Canada. But what matters more to her is the fact that her colleagues and students believe in her.

There is no doubt that Molly is a star. She has excelled in research, teaching, and business. Her career trajectory is emblematic of innovation. Molly has always been determined to build a better future, and given that she is hard at work with her team in the lab, the future holds great promise.

18

RUSSELL KERR

Culturing Microbes to Cure Infections

Charlottetown, PEI

Can we find natural products that will fight infections and treat cancer in humans? Russell Kerr thinks so and is seeking them in the ocean.

Born in Edinburgh, Scotland, in 1959, Russell came from an academic family. His dad taught in the field of medicine with one foot in Canada and the other in Scotland, while his mother taught art in elementary school. In 1969, the family moved to Burlington, Ontario, where his father helped establish a new approach to teaching medicine at McMaster University in Hamilton, Ontario. In 1974, they returned to Edinburgh, where Russell continued his education. He liked art and ceramics, taking after his mother, but he also enjoyed the sciences, resembling his father. As a teenager he had a paper route and worked a summer in the gravel pits. In 1977, when Russell was entering his final year of high school, his parents moved to Calgary, leaving him in a flat by himself. He says it was not hard for him, after all there were fifty pubs within a five-minute walk. Despite occasionally imbibing, he was a serious student.

He knew he wanted to study sciences, but he was not sure which field. If he went to university in Edinburgh, he would have to select biology or another field immediately. On the other hand, if he went to university in Calgary, he would be able to complete a first year in general sciences before deciding. He admits that Banff and the mountains may also have figured into his decision.

In 1978, Russell went to the University of Calgary, where he discovered organic chemistry. He was fascinated by the way molecules could be synthesized and wanted to learn more about new methods to make chemical structures. He was invited to work in a lab as a summer researcher. When he arrived, the project was all set up, and he synthesized what turned out to be a new chemical structure. Unsurprisingly, Russell was "hooked that first day in the lab." Looking back, he suspects that his professor had prepared his discovery, and "set him up" to succeed. Whether it was serendipitous or planned, that discovery ensured that Russell stayed in Calgary, where he did his BSc honours and completed his PhD in 1987. His thesis involved using synthetic methodology that could have practical applications in health care.

Russell did a postdoc at Stanford University in California with Carl Djerassi, a pharmaceutical chemist, known for his groundbreaking work in synthesizing antibiotics as well as hormones such as cortisone and norethindrone, the first birth control pill. Besides being well published, Djerassi was dedicated to the commercialization of research. Russell said the first thing he learned from him was to "think big." Russell has retained this advice throughout his career, always trying to address significant problems and welcoming entrepreneurship. He also learned that he should not be afraid to step away from his experiment in order to crystallize his thoughts and the process. When Djerassi asked him to set up a new lab at Stanford's Hopkins Marine science campus at Monterey, Russell was sent there on his own with one technician and they stayed eighteen months. The small campus was beautiful, located by the sea. The bicycle ride there in the morning was "stunning," and the warm weather was an added bonus. He learned how to scuba dive as part of his fieldwork, an ability that served him well in later life.

When he completed his extended postdoc in 1991, he wanted to find a university with a strong research and graduate program that was located on a coastline where there was good biodiversity, including corals and sponges. In 1991, he was offered a position at Florida Atlantic University, which met all the criteria. For more than fifteen years, Russell studied the structures and bioactivity of natural products from corals and sponges. Since this work had not been done before, he innovatively made the tools required to analyze them and mapped the metabolic steps of the natural process they used. Natural products derived from coral are well known to exhibit anticancer activity with great potential to become lifesaving drugs. While other labs had discovered the anti-inflammatory nature of natural coral products, Russell, working with scientists from Germany, elucidated the biosynthetic pathway, purified a key biosynthetic enzyme, and sequenced the gene. Now, with chemical synthesis tools, they are on the verge of being able to produce a final product, which is an effective anti-inflammatory. This, however, is not the end of their work, as it must be produced on a kilogram scale. Russell believes they are close and will succeed in the next year or two.

In 2006, Russell went to the University of Prince Edward Island (UPEI) as a Tier I Canada Research Chair. He says that there are many advantages to working in a small university with good infrastructure, and perhaps even greater possibilities for funding than in a larger institution.

Russell has spun companies out of his lab in Florida and in Prince Edward Island. One company, Nautilus, works on the microbiology of corals, which are rich microbial communities. Nautilus collaborated with Croda, the largest chemical company in the UK, and thanks to investments from the Atlantic Canada Opportunities Agency (ACOA) and the Atlantic Innovation Fund (AIF), Russell was able to negotiate the acquisition of Nautilus by Croda and the establishment of a branch of Croda on PEI.

Russell's work has evolved, and he has developed many collaborations around the world that enabled him to build an extensive collection of marine microbes. The resulting microbial library is a novel and essential resource with bacteria and fungi from the Bahamas, Colombia, Brazil, the US, the Middle East, the Dead Sea, Eastern Canada, and the Arctic. Russell notes that these interesting collections include new strains of bacteria and fungi from locations such as Frobisher Bay. This library is one of the most biodiverse collections in the world with over eight thousand unique microbes, which are stored in vials at –80 degrees Celsius in freezers that are located in different buildings for safety. Copies have also been sent to England and France. You can take a vial, start a culture, and then restock; so the resource should be self-sustaining for future research. Russell notes that these samples are very small and collecting them has almost no impact on the environment. In his academic lab, Russell and his students look for and identify new chemical structures. His Nautilus lab is primarily interested in creating new products, including antifungals, for use in personal care products and crop protection.

Among Russell's innovations are new ways to ferment bacteria to produce natural products. He says he is working in a "wonderland of new chemistry," where it is possible, thanks to metabolomics (the study of the chemical identities that remain from cellular processes), to discover new antifungals that he can then license to Croda.

Russell has also undertaken a project to "culture the unculturable." This project is designed to address the critical challenge in microbiology that only 1 percent of microbes from any environmental sample can be cultured. Given the tremendous benefits of drugs discovered from bacteria and fungi, accessing some of the "missing" 99 percent could prove to be extremely exciting. Marine sponges and corals contain highly bioactive natural matter produced by currently unknown microbes. These invertebrates represent a seminal challenge in the development of new microbe-culturing methods. The vast majority of microbes in sponges and corals have resisted all attempts to be cultivated. However, Russell's lab developed a sponge isolation chip, a device that has led to the isolation of new bacteria, one of which has proven to be a producer of a new derivative of a common amino acid that exhibits promising biological activity. More recently, he has developed a new tool and process involving the culture of sponge-derived bacteria in the lab using the kind of sponge from which the bacteria were originally extracted. Early signs suggest that this new process may prove to be highly effective in the isolation of previously uncultured bacteria.

Russell realized that the problem in "culturing the unculturable" was likely due to the need for an intermediate "domestication" period prior to culturing in an artificial environment in a lab setting. He consulted with his team, and they slowly developed the process.

Russell says his new microbe isolation methods will provide access to some of the 99 percent of inaccessible microbes and will permit the discovery of new and exciting natural products and enzymes. He adds that 80 percent of the antibiotics that are currently used as drugs come from only 1 percent of the culturable microbes. This means "the sky is without limit," if methods can be developed to culture a significant portion of the "missing" 99 percent.

Russell shares the advice he gave his son, who is just starting his research career in physics at Dalhousie University in Halifax: "Find something you love and work your tail off. Think big and do not only try to solve small problems. You have thirty to forty years ahead of you to try to make as big an impact as you can."

PART THREE

Innovatively Saving the Planet

Floods, forest fires, and dangerous storms offer stark evidence of climate change. Nations around the world regularly gather and set goals to reduce the consumption of energy and the release of carbon into the atmosphere, and every year we fall short of the targets. The automotive industry is racing to adopt battery-run vehicles, but they constitute only a fraction of the carbon emissions produced by other industries and manufacturing. Entire islands of debris float about the oceans, endangering the flora and fauna, not to mention passing vessels. We have even polluted space with debris that risks falling from the skies.

Yet, truly brilliant people are working to change and improve our world, and their innovative work offers us hope for the future. We begin with the story of a philosopher who realized that we cannot even talk about the decline in diversity of life on Earth unless we can first define the problem and agree on descriptions that allow us to measure change.

The preservation of natural resources and clean water is of foremost importance. A little-known but extremely impressive research facility in Canada is composed of fifty-eight lakes, where teams of researchers from around the world study the effects of toxicity and the means to remediate them. Canada is fortunate to have three oceans where plankton serve not only as food for fish but also sink more carbon than all the forests on the planet combined. The movement of fish around the globe and the ways fish, fisheries, and marine travel and transportation can coexist are crucial to our existence. Water also plays an important role far beneath the Earth's surface. A most extraordinary study involves the water trapped seven hundred kilometres underground in ringwoodite, which, like super-deep diamonds, is a high-pressure phase of magnesium silicate. Far beneath the Earth's crust, water can trigger some of the worst earthquakes. Our North provides a fertile ground for learning about these diamonds, and Canada is fortunate to be home to some of the world's leading research in the field.

We can also save the planet by finding new energy sources, creating communities that are totally off the grid, and exploring nuclear fusion, to name but one area that is intriguing scientists and investors alike because of its potential to solve the entire world's energy crisis with a single plant. The race to produce this energy is underway and hope is on the horizon.

The eyes of astronauts focus on preserving the most spectacular and pristine places that remain on Earth, while back on Earth, imaginative researchers work on the technology required to send people into space and are, at the same time, inventing ways to collect space debris, which could damage the Earth or collide with space platforms or rockets.

Our most precious resource is talent and we must ensure that future generations of researchers have scientific knowledge, a passion for the environment, and respect for the people who dwell on the land they have inherited and for the traditional knowledge they possess. Sharing knowledge is perhaps the most important way we can contribute to the future. In Canada we are fortunate to have some extraordinarily talented, wise, and generous teachers and mentors. With the inspiration they share, along with the significant discoveries, bold innovations, and the many large and small victories we see across the country every day, we just might win the race to save our planet.

19

FRÉDÉRIC BOUCHARD

Merging Philosophy and Science

Montréal, QC

Freedom was one of the greatest gifts Frédéric Bouchard's parents gave him. They allowed him to follow his natural curiosity, to experiment, and discover. Years later, when a professor said, "In philosophy, you have the right to ask anything, to question everything," the die was cast. Frédéric decided then and there to pursue his university degree in philosophy. However, he had always considered philosophy and science together as parts of a whole. He attributes this realization to having read while in elementary school a biography of the artist, scientist, and philosopher Leonardo da Vinci.

Born in Beauport, Québec, in 1976, Frédéric grew up in Montréal. As a young child, he experimented with small electric motors and built an automatic lamp with motion sensors that would turn it on and off. Although it burst into flames, Frédéric's parents encouraged him to pursue his passion, while gently suggesting that he should make sure to seek as much knowledge as possible prior to launching an experiment.

It was the freedom to explore, to satisfy his curiosity, that encouraged young Frédéric to enter a citywide elementary school invention competition. He designed and built a model of something he wished for and something we have all desired at some point in our busy lives: a self-making bed! He won the top prize.

Frédéric's scientific adventures were nourished by the children's science magazine *Les débrouillards*. With his friends, he eagerly awaited each issue and tried nearly all the experiments in the magazine. One day, they noticed that the magazine was produced close to their school in Montréal. The small group of serious youngsters decided to visit the editorial offices. They arrived unannounced and were well received by the editorial staff. After speaking with the writers, artists, and editor, each was given a sketch, and Frédéric was later invited as a guest youth journalist, honing, at an early age, his communications skills.

Growing up surrounded by books, Frédéric's intellectual pursuits were encouraged by his parents and teachers during his early years at school, and his fascination with technology and philosophy took root thanks to the freedom he had to explore different topics. After completing CEGEP, Frédéric went on to earn his BA (1997) and MA (1999) at the Université de Montréal. During his studies, he took courses in various fields, including neurobiology for curiosity's sake, and today he encourages students to do the same. Wanting to expand his horizons beyond Québec and Canada, he would go on to Duke University in North Carolina for his doctoral work (1999–2004), and then to the University of Toronto (2004–2005) for a postdoc, after which he returned to the Université de Montréal as a professor, soon becoming directeur du centre interuniversitaire de recherche sur la science et la technologie.

Frédéric's innovative mind was evident when he wrote his PhD thesis in philosophy inspired by science and biology for his doctoral degree at Duke University in 2004. His research looked at the theoretical foundations of evolutionary biology and ecology and the relationship between science and society. Frédéric studied the ways biologists define and measure the health and fitness of biological species and their impact on the evolution of ecosystems. He suggested a new definition of evolutionary fitness after observing that the textbook definition was not what scientists actually observed in many biological cases such as clonal species or symbiotic communities. The evolution of species is foreshadowed by and includes changes to behaviour and reproduction. Frédéric realized that something was amiss: either the evolutionary phenomena were not described appropriately, or the definitions needed to be revised. Frédéric's contribution to environmental science is fundamental. We cannot determine if there have been evolutionary changes if we have not defined the problem and described the criteria for determining the nature of the changes.

Not surprisingly, Frédéric's favourite philosopher is Charles Darwin. When people protest that Darwin is really a scientist, Frédéric recommends that they read *On the Origin of Species.* It becomes clear when reading the book that Darwin was both a scientist and a philosopher. In the seventeenth, eighteenth, and nineteenth centuries, all philosophers were scientists and all scientists were philosophers. It is evident that Frédéric is a true descendant of the Age of Enlightenment.

Today, he still combines the two fields, speaking to biologists about the power of philosophical analysis to improve scientific practice and sharing with philosophers about how the wonders of biological processes can help elucidate classical philosophical problems. With colleagues from Duke, Cambridge, Toronto, Paris 1 Panthéon-Sorbonne, and the Université de Montréal, he founded the Consortium for the History and Philosophy of Biology, and as the current dean of the Faculty of Arts and Sciences at Université de Montréal, he helps students and colleagues explore their own multidisciplinary interests.

As chair of the digital publishing platform Érudit, and a member of the board of Mila (the Québec Artificial Intelligence Institute) and the Institut de recherche en biologie végétale (Research Institute in Plant Biology), he shows the value of mixing disciplines and broadening the public's understanding of scientific research. Érudit makes academic publications available to everyone, whether they have access to a library or not. He believes Canadians have tremendous talent, but wishes they had more opportunity to expand their horizons, be it by following their academic passions, by learning from the past, or by expanding their view of the world through travel. Study abroad at Duke and a stint as invited professor at the University of Lisbon in 2013 changed Frédéric's life, and he sees that travel is one way to stretch one's mind, to acquire knowledge, and to establish greater solidarity with humankind. However, he adds that there are many ways of travelling: learning Latin or ancient Greek is like travelling through time and is equally enriching.

Frédéric feels an ethical sense of responsibility to future generations. He taught a course on sustainability and the environment and recalls with pleasure bringing together students from different backgrounds, cultures, and disciplines who felt initially that they had nothing in common, but ended up realizing how much they had to share, starting with a thirst for knowledge and a deep understanding that each person merits respect and dignity. Contributing to universities and teaching students is Frédéric's response to the economic and political stratification that are prevalent in the world today. In addition, he is committed to helping institutions that increase society's research capacity over time. He thinks about how these institutions plant the seeds of knowledge and the passion for learning that will grow into the ideas that will make the world a better place in two, five, ten, and even a hundred years from now.

As a personal pandemic project, Frédéric decided to try woodworking and learned antique joinery. To complete one particular project, he needed a molding plane, which is no longer manufactured. He found one in an antiques shop. While he was restoring it, he noticed the maker's mark on the plane, and after some sleuthing he learned that the tool had been made in Roxton Pond, Québec, in the nineteenth century. This, of course, led him to learn about life in Roxton Pond in 1865, when Mr. Sem Dalpé bought a furniture store. The previous owners had specialized in making wood finishing planes, and Dalpé, seeing an opportunity, started what was to become the largest tool-producing center in nineteenth-century Canada. Through this hobby project, Frédéric acquired a concrete appreciation for the momentous change that the Industrial Revolution introduced to communities, and how technology alters the way human beings conceive of their own place in the universe. It has also led the ever open-minded and thoughtful Frédéric to contemplate the creation of a new course that would task students with a physical project like building a chair or a robot before tackling philosophy. The idea is to improve their thinking while working at the intersection of different fields and while using their hands as well as their minds. One can only build a better world with the mind when one has an appreciation for the diversity of human experience.

Driven by his voracious appetite for understanding how everything works, Frédéric has himself designed various mechanical devices. His ingenuity is to discover the value of the past, to make us question the present, and adopt a sustainable and ethical framework for the future.

20

RICHARD FLORIZONE, MATTHEW MCCANDLESS, AND VINCENT PALACE

Fifty-Eight Ordinary Lakes and Three Wise Men Solving Environmental Challenges

Saskatoon, SK; Winnipeg, MB

Fifty-eight small lakes in a sparsely populated area of Northwest Ontario constitute a natural lab that, according to its website, permits research at "the level of the entire ecosystem, improving our understanding of human impacts on the environment, influencing policy, environmental best practices and supporting public awareness."

These lakes are special because they are *not* unique. They represent an entire global ecosystem, and through the scientific research being conducted, they are revealing secrets of global significance.

The fifty-eight lakes are managed by the International Institute for Sustainable Development (IISD) under an agreement with the government of Canada and are located on Anishinaabe Treaty 3 land, where scientists work side by side with elders and members of the Indigenous community. The lakes are three hundred kilometres east of Winnipeg, Manitoba, and one hundred kilometres west of Dryden, Ontario. Scientists have been working at the lakes for the last five decades on topics that include acid rain, algae blooms, and the impact of mercury on the flora and fauna. Vince Palace reminds us that today they stand on the shoulders of giants, scientists like David Schindler, the renowned Canadian ecologist, who was among the first to ensure the area was protected.

Research conducted at the lakes allows scientists to observe the effects of pollution on the water, fish, and plants, and to understand the amount of time it takes for nature to reverse the effects of toxicity. Over 250 researchers from across Canada and around the world participate in this work, along with the team of 40 scientists at the facility. The challenge is, Richard Florizone states, and they all agree, "too big for one individual or one country. We must find ways to build new collaborations and challenge others to join us in this work. Water is the driver of life. While the lakes are not unique, the facilities to study them are world-class and the only such facilities in the world."

This is the story of three extraordinary and innovative people who have come together to lend their unique talents to the development of an extraordinary resource for environmental research.

Richard Florizone, past president and CEO of IISD, was born in Prince Albert, Saskatchewan, in 1967 and inherited a love of education from his parents and grandparents. They taught him respect for science and gave him an understanding of the natural world.

In 1990, he received his bachelor's degree in engineering-physics, followed in 1992 by his master's in physics, from the University of Saskatchewan (USask). He was awarded a Natural Sciences and Engineering Research Council (NSERC) award, which allowed him to do summer research at USask's accelerator tube that created energy through particle acceleration. He then went on, with the guidance of Dennis Whyte, a world expert in the development of magnetic fusion energy systems, to do a PhD in physics (1998) at MIT.

His studies in engineering and physics, combined with an innate talent for finance and a set of strong leadership skills, have led Richard to a career that included stints at Bombardier, the International Finance Corp., the World Bank, Boston Consulting Group, and the Quantum Valley Ideas Lab. He served as president of Dalhousie University in Halifax, where he helped found the Ocean Supercluster and the Ocean Frontier Institute that built links between businesses and researchers to protect ocean resources while growing the economy. He is senior executive fellow at the Waterloo Institute for Sustainable Energy and fellow of the Canadian Academy of Engineering. He has won awards for his work on the environment, including the Clean50 Award, and in the second quarter of 2023 became the advisor to World Energy GH_2, developing Canada's first commercial green, hydrogen production.

Winnipeg-born **Vince Palace** is an aquatic toxicologist with twenty-five years' experience on projects studying the impact of hydroelectric power, oil, gas, and mining on aquatic ecosystems, and currently is head research scientist at IISD. He grew up interested in science and sports, and today those interests continue to converge. In 1987, he completed a three-year general science degree at the University of Manitoba. When he realized that he should have done a four-year degree, he returned as a special student and then continued directly to graduate studies. His graduate research tested the effects of cadmium on aquatic species. While working on his master's degree, he had hoped to publish just one chapter of his thesis, but to his surprise and delight all four were published. In 1996, he completed a PhD in zoology at the University of Manitoba and has since enjoyed a fine career as an adjunct professor and lecturer in biological sciences and environmental geology at his triple alma mater and is also an adjunct at Lakehead University in Thunder Bay, Ontario, and the University of Saskatchewan in Saskatoon.

Matt McCandless, both senior director of fresh water and executive director of IISD, was born in Winnipeg, and "decided at the age of twenty to go as far away as possible from Manitoba." He travelled the world, spending time in Africa, Asia, and the South Pacific. He ended up in Australia, driving forklifts in Sydney. One day, contemplating the beautiful countryside, he thought that he could spend his entire life there. But he also thought about driving forklifts for the rest of his days and realized that he would prefer to make a difference, to do something more impactful.

He recalls that he had previously volunteered on an organic farm and later worked at a commercial orchard, where the apples had a filmy covering from the pesticides. Despite frequent spraying, there were still bugs in the apples and if you held one in your hand and then touched your face or eyes, they burned. He also remembered a small hydro plant in a village in

Asia where there was barely any water flowing in the stream. The project failed because the engineers had not visited the site or consulted the local population. They could have saved at least a million dollars had they done so. He saw too many failed projects and dashed dreams in his travels.

He therefore decided to return to Canada and study biosystems engineering at Dalhousie University. On graduation in 2001, he became senior engineer in a consulting firm, where he was encouraged to write and think about people and the impact of their energy consumption on the environment. This work became part of his thesis on rural electrification, for which he earned his master's degree in 2007 in natural resources management from the University of Manitoba. He returned to the University of Manitoba to do a PhD in biosystems engineering, which he completed in 2018. It included a study of water, energy, and agriculture. He has completed dozens of projects on water, agriculture, and bioindustrial development in Canada, Africa, Asia, and Latin America. He continues to research ways to optimize land use to protect the water and wetlands and contribute positively to the health of the population and the environment. He joined IISD as associate vice president of water and managing director in 2006 and has served as executive director of IISD-ELA (Experimental Lakes Area) since 2014.

Combing their expertise and broad experience in research, management, and leadership in bringing the public and private sectors together both nationally and internationally, this powerful team has made extraordinary contributions to our knowledge and ability to mitigate the impact of pollution in freshwater lakes. They are studying the effects of pharmaceuticals and microplastics on the environment. They have also researched oil spills that may affect fresh and salt water, for example. During the *Exxon Valdez* disaster, attempts to remove the oil using harsh methods caused additional damage to the environment. The IISD-ELA team is now asking what would happen if the oil were left to degrade in place. How long would it take and what damage would it cause? It is doing the research and assessing the risks and benefits of each approach to mitigating damage to the environment, not only in the case of oil spills but also of the chemicals that pollute the air, land, and waters.

Richard says, "Look at the climate crisis. We picked a war with nature and must reverse the situation. For generations people have only thought in the short term. We need to adopt what the Indigenous peoples have always known and learn how to end this war and repair the damage that has been done."

Matt adds, "We are part of nature. If we do not take nature seriously, we will lose our planetary balance. What we do here has implications for the world. Take, for example, the seven great lakes of Africa. In Canada if the lake is polluted, cottage prices go down. In Africa, millions have no way to earn a living."

Vince concludes that this is the theme of all the projects at the ELA. In addition to being a research organization, they undertake outreach and education. They include communities and try to harness the power of citizen scientists. People have a passion for the environment, and there is hope that a myriad of projects, large and small, will make a difference here and around the world. They collect data, share it, and want people to own and use it. They are introducing tools to communities so they can, using sophisticated techniques, better identify stressors to the environment.

IISD has resulted in publications in *Science* and *Nature*. Their work contributed to the Minamata Convention on Mercury, limiting global production and outputs of the dangerous heavy metals. Their topic is not only the environment but human life and existence. The recipe for success is global excellence and local relevance. As Richard said at the beginning, "Nobody does anything alone." And he adds that he is just the conductor of a small orchestra. For the future of our planet, let us hope that it will grow to an even larger, international symphonic orchestra and that Vince and Matt will lead their sections with bravado and precision, creating a new hymn to the capacity of humankind to reverse history and to end the war on nature.

21

RICHARD DAVIS

Inspired by Science and Jacques Cousteau

Halifax, NS

Born in 1961 and raised in El Paso, Texas, desert-dry, cactus country, it is somewhat surprising that Richard Davis made the decision to become an oceanographer. But he had become fascinated early on by sharks, whales, and all marine life watching *The Undersea World of Jacques Cousteau* on television. And his favourite teacher, who taught seventh-grade science, really cared about the students, encouraging them to explore their interests.

Along with his five siblings, he attended university with the moral support of their parents, who could not afford to contribute to their education financially. For this reason, Richard always had one or two part-time jobs during the school year and full-time jobs over the summers. He learned something useful in each job. For example, working the night shift as a janitor in the Texas Department of Public Safety, he learned how to get along with many different people, and at McDonald's he perfected his multitasking skills. As a main well logger at the Bureau of Economic Geology, he transcribed and coded information every day concerning the offshore oil wells, observing at the same time how an office might function more efficiently and learning a great deal about marine oil wells.

In 1981, he went to the University of Texas at Austin to do an undergraduate degree in aquatic biology, since the institution did not offer oceanography.

In 1986, Richard subsequently completed his master's degree in biological oceanography at the University of Texas, which had a marine biology station at Port Aransas, where he spent two years doing research. His thesis was on the effect of turbid water on marine plants. He later conducted a similar study in the Antarctic looking at ozone depletion and phytoplankton, finding that they were likely not seriously impacted due to vertical mixing in the upper layers of the ocean, which means the phytoplankton are generally not in irradiated waters long enough to be deleteriously affected. Richard jokingly says that he wrote a "seminal study on the topic that only twenty people understood." However, the ocean science faculty at the University of Washington in Seattle had read it, and not only understood it, but also valued his work, because they invited him there for graduate studies. That summer, he took a course in bio-optics and courses in physics and biology at Friday Harbor, the marine biology field laboratories at the University of Washington, and made a graceful transition to Seattle. Several years later, he was nearing the finish line for his PhD, when he received an offer for a position with the National Oceanic and Atmospheric Administration (NOAA), a US agency that forecasts weather, monitors climate change, participates in coastal reparation, and supports marine commerce. The lure of the sea and spending time on coastal vessels was impossible to resist.

Soon Richard was modelling the pollack fisheries along the coast of Alaska and the Bering Sea, which were collapsing at the time. In the 1980s walleye pollack was reputedly the most abundant and lucrative natural fishery in North America. However, the "Donut Hole" stock in the Aleutian Basin of the central Bering Sea, an international zone, was overfished by large fleets of factory ships belonging mainly to the US and the USSR. There was also a lack of knowledge of the biocomplexity of the fish population and the best means to harvest. This would be one of the worst collapses in fisheries in North American history. Richard also set up observation and weather stations on some of the tiny islands, uninhabited except for the bears and other wildlife. He would be flown in by the Coast Guard in a helicopter from Anchorage to set up the equipment that would transmit data to scientists onshore. NOAA also put out buoys called "Peggy buoys," after a woman who read the weather on the radio on the West Coast for many years. Richard persuaded NOAA to let him load some extra equipment in the base of the buoys, which were themselves anchored in the sea and used to transmit signals about the winds, waves, and weather. That year the water froze before the buoys could be removed and the ice sheared off their surface, but the result was that Richard had made the very first-ever under-ice observation of phytoplankton fluorescence. Richard's research goal, emanating from his doctoral work at the University of Washington, was to study the bioplankton, a basic food source for marine fauna and of oxygen for the Earth. Richard's research on this topic was indirectly related to the pollack study, and he worked on both at the same time.

While Richard was at NOAA, his advisor, Dr. John Cullen, who was an oceanographer specializing in the study of plankton at the University of Washington, took up a position as professor at Dalhousie University in Nova Scotia, and asked Richard to come to work with him there. Richard agreed and was offered a research and technical support role in the Department of Oceanography in 1996, which is how he ended up on the Atlantic coast, where he says he "just did what was necessary to make research projects work." Richard worked with Dr. Cullen for eighteen years, setting up and doing experiments, and co-authoring numerous papers. It also meant taking on contracts to study diverse fish stocks not only in Canada but in Greenland.

Richard also set up the first real-time automated research buoy in Canada. It would provide weather information to receivers on the shore. There was also a plan to create a Centre for Marine Environmental Prediction for weather conditions in Lunenburg, Nova Scotia. Richard got this up and running—yet another feat! Thirty years later, the buoys and sensors Richard built have been moved to Bedford Basin, a large, enclosed bay at the end of Halifax Harbour. Now operated by Fisheries and Oceans Canada, they report valuable data to be employed in developing new methodologies and ideas for safely and economically removing carbon from the atmosphere. Rather than report the weather, they now report information that will determine how to create conditions to improve the air quality.

Between 1992 and 1997, Richard returned for brief periods to Alaska, where he was able to fly up along the Bering Sea and study the heterogeneity of the bioplankton concentration. NOAA had aircraft specifically equipped with sensors, called Hurricane Hunters, that flew directly into the eye of storms to collect data. During the off-season the planes could be used for other scientific work. Richard thus was able to fly to Alaska. He also spent three seasons cruising the Antarctic and lived for three months at Palmer Station, studying the effects of light on plankton in the Antarctic. This work was, and remains, important because phytoplankton are the foundation of the aquatic food chain, and everything from microscopic zooplankton to whales depend on them to survive. In addition, while they provide food for fish, plankton produce the oxygen we breathe, sink carbon at the bottom of the sea, and can also be used to create biofuel. Thinking back on the challenges of these expeditions, Richard quotes the old saying, "You go to Antarctica once for adventure, the second time for money, and the third because you do not fit in anywhere else." This was surely not the case for Richard, who had found his place in Halifax and at the university and the Ocean Tracking Network (OTN), which runs a number of mini autonomous submarines to track fish and pollutants in all the world's oceans. The OTN tracks whales around the globe and can interpret their activities at a distance, recording their sounds. It also studies fish populations and the movement of crustaceans. The OTN tags sharks and studies their behaviour, as it is influenced by the availability of food, the temperature of the water, and the activities of large ships plying the seas.

The OTN's mini submarines, called gliders, have gone more than three times around the globe. They are slow-moving, at one kilometre per hour, so the trackers have to be "very patient people." There are twelve to fifteen gliders and five to eight of them are operating at any one time. The research is client-driven, with, for example, ecologists, or companies involved in harvesting fish, or governments working to capture and sequester carbon and to regulate fish-harvesting activities. All seek scientific evidence as they determine when and where to fish and if they should. Richard figures out how to provide the equipment they will need for their projects and makes sure it is functioning. A project to listen for North Atlantic right whales might require the glider to remain at sea for four months in all weather conditions. Richard has developed the sensors that distinguish one species of whale from another. However, he points out that the gliders can only detect whales when they are vocalizing, and they appear to vocalize only when feeding and not when transiting. Recently, there have been requests for carbon sensors, and some are located at the water's surface, while other sensors are placed on the ocean floor. Each mission has a different profile.

Richard has also created different kinds of incubators that make it possible to study the physiology of phytoplankton. He has worked out ways to screen the phytoplankton and is ruminating about the possibility that their excess energy could be changed to biofuel, an attractive alternative given the price of oil. He is also thinking about ways to deal with the phytoplankton that come in bilge water and are discharged into bays and harbours, polluting coastal waters and endangering native species.

His advice to those contemplating a career as an oceanographic researcher is the same he has given his daughter: "At the end of the day, if you did not enjoy yourself, you were working. If you did enjoy it, you found a career calling." He says to find something that gets you up in the morning, and cautions that one should not become a "wage slave." For those with a passion and interest for it, study STEM (science, technology, engineering, and mathematics). Computer science and math are important in all fields. And finally, be sure to pass on what you learn to others.

Richard knows the sea and marine life. He asks profound questions, enables good science, and collaborates with other researchers to achieve the proper balance between service to the economy and preservation of the marine ecology. And that is exactly what innovation can achieve: making goals attainable and the end product more efficient, economical, and beneficial. And that is indeed beautiful.

22

D. GRAHAM PEARSON

The Treasure Hunter

Edmonton, AB

For D. Graham Pearson, diamonds are his life, passion, and work. A diamond is at once "a wannabe graphite" and "the most remarkable of any mineral."

His story begins in 1966 in Pontefract, a small town in Yorkshire, England, in a family with a strong work ethic. Graham understood from a young age that "things are not always easy in life." As a lad, he would typically have an early start to the day, delivering milk from 4:30 to 7:30 a.m., six days a week. Then he went to school. He liked his work and fondly recalls the "cream chickadees," the cream that separated, rising to the top of non-homogenized milk, freezing on cold days into a stick that peeked up over the cap of the bottle.

His parents did not pressure him and his twin brother to study. Nonetheless, their father noted that if they did *not* study, they would "end up in the pit (the coal mines) like anyone else," and supplied them with books and encyclopedias, which Graham loved and, to this day, considers a great investment.

His father was a safety officer in a chemical factory, occasionally performing demonstrations to help illustrate the dangers of the chemicals. Each year he would toss ten-pound blocks of metallic sodium into the canal, which would immediately burst into blue flames, convincing the workforce of the danger. So it is not surprising that by the age of eight, Graham was already fascinated by chemistry and he even began thinking about the possibilities of science. He wondered, at that young age, for example, how the world could save energy.

Graham and his brother were the first in their family to go to the university and were able to do so thanks to government grants. Graham determined that he would be a mining geologist, and so he went to the Royal School of Mines at Imperial College London, where he received a first-class degree in geology. There were few opportunities for employment as a geologist specializing in mining in the UK, and he liked research and working in labs, so he went to graduate school at Leeds University, where he says there was "an amazing project on diamonds—the only one in the United Kingdom!"

He then had the good fortune to do interesting fieldwork collecting rocks in Morocco for three months. He and his field assistant arrived by donkey in a mountainous region where his advisor, Peter Nixon, professor of earth sciences, had stumbled on a unique suite of rocks containing graphite, which, along with charcoal and diamonds, is an allotrope, or different form of carbon. Graphite is a lower-pressure form of diamonds, so its presence proved that the Earth had shoved up from its mantle, the layer of hot, solid rock between its crust and the core that had previously been diamonds when under greater pressure deep below the surface.

After a postdoc from 1990 to 1994 at the Open University in Milton Keynes, England, and the Carnegie Institution for Science in Washington, DC, and some fifteen years as a professor at Durham University in England, he was recruited by the University of Alberta in 2010. He felt that it was a tremendous opportunity. He knew many Canadian geologists and had worked in the Arctic. The University of Alberta had the largest chemical lab in Canada. He says, "One needs big equipment to do innovative science," and he feels that the equipment he was offered compares favourably with that found in labs around the world, making his lab "perhaps the largest geochemical lab in Canada. It enables the measurement of tiny amounts of geomaterials that are extremely difficult to measure."

While much of Graham's work is lab-oriented, he disappears frequently in the field for two months at a time on a mission to unlock the secrets of the early Earth, studying how the Earth was formed and noting how hot, liquid rock rises to the surface and continues to evolve. He often found a few very old diamonds along the way. Graham modestly attributes some of his finds to serendipity and others to extreme exercises in patience, but he cannot deny that he created a new method for determining the age of single diamonds. He was also the first to develop a method to analyze and quantify trace elements in diamonds—a method that enables scientists to identify the exact origin of a diamond. This information is used in the Kimberley Process, which is employed by administrations, industry, and civil societies from around the world to prevent the trade of rough diamonds, which have been used to fund wars and rebel movements in zones of conflict.

His contributions to the field are numerous, but the discovery of water in the deep Earth has certainly had the biggest impact. Theoreticians had long speculated that a key mineral, ringwoodite, a high-pressure variety of peridot, had a structure that enabled it to capture and hold 1.5 to 2 percent of its weight in water. Researchers verified this speculation using high-pressure experiments. There was, however, no natural sample—and thus no proof. For over twenty years researchers had been looking for a sample of the mineral, whose existence would demonstrate the theory. Just about everyone had given up the search. This is where Graham's ringwoodite adventure begins.

He decided to look at the subclass of diamonds known as super-deep diamonds. Diamonds are usually 150 to 200 kilometres beneath the Earth's surface. Super-deep diamonds are 700 kilometres deep in the Earth. Like super-deep diamonds, ringwoodite, a high-pressure phase of magnesium silicate, is formed in the Earth's mantle at a depth of 525 to 660 kilometres. Graham speculated that he could perhaps find ringwoodite along with these diamonds.

The discovery was made, according to Graham, through a combination of luck, patience, and knowledge. His group had been studying many diamonds and were running out of funding to continue their work. In 2014, Graham led an international team to collect samples in Brazil.

They decided to look at a specific class of diamonds that were the "grubbiest of diamonds" and extremely small: three millimetres, with inclusions of thirty microns in size. The samples were irregular and rather fractured, demanding a high-quality spectrometer to measure the water contained in the ringwoodite they hoped to find. The water in the ringwoodite is structurally bound to it. Graham explains that it is like a fizzy drink, where carbon dioxide dissolves in water. At high pressure, the water is trapped in the mineral. Much of our planet's deep water is contained in ringwoodite and related minerals, likely transported there by plate tectonics. With colleagues at Carnegie Science, Graham speculated that the recycled water may play a crucial role in triggering deep-focus earthquakes. At seven hundred kilometres under the surface, rocks are supposed to behave plastically, but fluids likely trigger different behaviour.

In the lab at the University of Alberta, Graham and his doctoral student, John McNeill, began working on the samples they had collected. They focused on the preparation of the samples and decided not to polish them, as the heat could possibly cause the minerals to revert to a different state or another form of the same element, just as the diamonds in Morocco had become graphite. They were "picky about the wavelengths of the lasers used to make measurements and the length of time [they] focused the laser beams on the tiny inclusion." The ringwoodite was far from its stability field and so the slightest thing could trigger its reversion to another lower-pressure mineral. The samples were irreplaceable. Graham says, "If you ruin them, it is game over." The very last sample they tested contained ringwoodite, confirming their findings and establishing the theory as fact. After two decades of research, the results speak for themselves. Graham and his student found the holy grail of geologists. They proved that water exists in the deep Earth and is contained in the rocks. Their results have prompted scientists to attempt to determine if the water was there when the Earth was formed or if it was transported down, in which case it may trigger deep-surface earthquakes. This information may indicate how earthquakes form and alert us to seismic hazards or eruptions.

Graham is passionate about diamonds. He says, "They are remarkable because they are the hardest mineral, the one with the highest thermal conductivity, they possess amazing semiconductor properties, meaning they have great potential in quantum computers, for example." He has just finished co-editing a nine-hundred-page volume entirely devoted to diamonds. His stunning discoveries have meant that he was named fellow of the Royal Society of London, England, and of the Royal Society of Canada, and he is the Canada Excellence Chair Laureate in Arctic Research at the University of Alberta.

Graham sometimes worries that he will run out of ideas, but each discovery leads to another, and he has no dearth of projects to which he wants to turn his attention. What if he could discover how earthquakes form? What if a Mars exploration mission could return with rock samples from its interior? Of course, passage to Mars is, well, just another problem to solve!

23

JIM WILLWERTH

Smart Agriculture and the "Best Reds"

St. Catharines, ON

Jim Willwerth loves nature and is dedicated to making our habitat sustainable while supporting agriculture and, in particular, viticulture—the cultivation and harvesting of grapes for winemaking. Along the way he has developed innovative processes and tools that have helped winegrowers ensure their vines survive despite temperature fluctuations, climate changes, and plant viruses.

Born in 1978, Jim grew up in Niagara Falls and Port Colborne, Ontario, where he learned about growing fruit and making wine from personal experience. The family cottage was on a farm, and as a child, Jim drove a tractor and learned how to handle agricultural equipment. He would accompany his mother to Niagara-on-the-Lake to pick strawberries and peaches, and he knew the local farmers, many of whom have since transitioned to growing grapes. Each fall, he helped his father, who bought and pressed grapes, making wine in his garage.

Jim's parents were one hundred percent in favour of education. His mother was a computer programmer, and his father a pharmacist, who studied math and chemistry, but was a philosopher at heart. His parents wanted their children to excel and would return from parent-teacher interviews encouraging Jim and his three siblings to do even better. Jim was always interested in nature and scored well in math and science. When he decided to pursue graduate work, his parents could not have been more pleased.

In 2001, he completed a bachelor of science degree in chemistry and biology at Saint Mary's University in Halifax, where his family had roots and relatives. He then returned to the Niagara region, where he volunteered with various not-for-profit organizations doing inventories of the different varieties of plants and working with species at risk of becoming extinct. During this time, he decided to do a postgraduate certificate program in grape and wine technology at Brock University in St. Catharines, Ontario. He completed the program in 2004, and then worked in a vineyard producing some of the "best reds" in Canada. He decided that he loved research and "the plant side more than the wine." He looked at the work being done in the US, but really liked what was happening at Brock University, and in 2005 he started graduate work in plant science, studying the terroir effect, which is the way water, climate, and soils affect the growth of the vines and the grapes, and ultimately, the wine. He says the amount of stress that the plant undergoes will affect the quality of the wine. He did three seasons of fieldwork to complete his thesis. During this time, he met a researcher, Dr. Amy Bowen, coincidentally from Nova Scotia, who became his wife.

"Things have a habit of coming full circle," Jim says, thinking of his now-double family ties to the Maritimes. His wife is research director at the Vineland Research and Innovation Centre, so talks around the dinner table at their home are more often than not about research and innovation.

In 2011, Jim got his PhD and became a staff scientist at Brock University's Cool Climate Oenology and Viticulture Institute (CCOVI), working half-time in the lab and half-time in the field with growers. He has recently become a full member of the faculty in the Department of Biological Sciences, adding teaching to his busy schedule. From his experience in the field, he understood what the growers needed and wanted to know. From his own studies and research, he also knew how to solve problems. For example, grape growers had invested in wind machines to protect plants in the winter, but growers needed guidance on when to run them. These machines are unfortunately very noisy (they sound like helicopters and were not popular with local residents), and they were also expensive to operate, costing hundreds of dollars to run. There were over six hundred machines in Ontario vineyards alone. Jim said the team at CCOVI had to figure out exactly when the machines should be turned on and in precisely which vineyards. This was a complex innovation. First, he used a cold hardiness monitoring program for grapevines (the largest in the world) along with a very precise weather tracking system with sensors in vineyards across the region. The large data sets they produced were incorporated into a web-based platform called VineAlert that was available to all growers. Growers now had access to close-to-real-time data to know exactly when the wind machines were needed based on temperature lows and the vines' ability to survive at those temperatures. Traditionally, wind machines had been installed to protect against frost, but in the Niagara region, midwinter conditions can create plummeting temperatures on the ground, which can cause severe damage to the vines. In addition to freeze-mitigation strategies, Jim also provided advice to vintners on when to prune and when to leave extra buds to compensate for damage if freeze injury did occur due to a polar vortex or other cold event. Economic analysis studies indicate that this work saved the growers more than $55 million over a five-year period by avoiding lost sales, saving on vine renewal and replanting costs, and reducing operational expenses.

With his colleagues, Jim will be utilizing this data to create cold hardiness models that will help determine how future climate predictions may affect grape suitability in Canada and how to expand the VineAlert program without additional sampling in the field. The next step will be "smart" wind machines that will self-learn based on the model and vineyard site conditions. Jim wants to develop these systems in collaboration with engineers using artificial intelligence to improve the sustainability of farming. These approaches could have an impact beyond grapes to save any perennial crop from cold events.

Jim notes that farmers are one of the most innovative groups in the world. They think outside the box by necessity. For example, in Québec some vintners were experimenting with geotextile materials to cover the vines during the winter. Jim's research again proved helpful in evaluating feasibility and the best placement of materials over the vines. He found that geotextiles worked better in that region than burying the vines with soils and mulch because the soils could wash away or lead the buds to rot.

The biggest risk associated with climate change is extreme and erratic weather. One way to address this potential threat to crops is to improve resiliency in plant matter and develop more mitigating strategies. Among many other projects, Jim is interested in grapevine cultivars, clones, and rootstocks that may be more suitable to climates in growing regions across Canada. He has been working on national initiatives concerning cold hardiness as well as grapevine evaluation programs. He helped set up technology for monitoring cold hardiness and is coordinating with labs across the country to collect data. In addition, since there is currently no national clean-plant program in Canada, he has projects with colleagues at Brock and across Canadian institutions to support the Canadian Grapevine Certification Network. Right now, the grape industry relies heavily on imports for certified plant material. Establishing a Canadian program would reduce the spread of viruses. It could also be used as a model for peaches, plums, and other fruits. Jim works closely with grapevine virologists and pest management specialists to integrate his grapevine evaluation research to support new diagnostic testing methods and to help reduce spread of disease.

Jim's advice is to "never stop learning. Think outside the box. Do not be afraid to ask questions. Have a good understanding of the subject. Do not have tunnel vision. Always explore more. Everything starts with fundamental research. Collaborations help, but be careful to keep yourself from going down a rabbit hole when the projects expand beyond the initial scope. Creative thinking is important, but one must keep one's focus and remember objectives."

When Jim gets to take a day off, he likes to go fishing for muskie (catch and release, of course). There is always work to do with vines and plants in his mini arboretum and woodlot, where he propagates plants from local sources, saving species at risk, improving biodiversity, and removing invasive species. He is passionate about our natural heritage. He suspects there is much to learn about the species present in our ecosystem and thinks that plants are largely underappreciated. Plants provide food, fuel, and medicine, but we understand very little about their uniqueness, and we are just starting to understand their interactions with other organisms (their ability to attract, repel, and communicate). He believes we need to learn more about aspects of plant life, such as their circadian rhythms and organismal interactions.

As Jim looks out his window, he says he simply "loves nature, which is really incredible." He also loves what he does, which certainly is incredible. He acknowledges there is still so much to learn and so much to do. We need more than one solution to the climate crisis. Remember, he says, "We have to feed the world."

24

ROSS KONINGSTEIN

Google's First Director of Engineering

Ottawa, ON; Stanford, CA

Ross Koningstein says that by no means was he a bookworm, even though as a child he would check out and read every single book in the library on whatever subject interested him, for example: flight and aircraft, the technology of radio controls, and telemetry. He was just naturally curious.

He also liked "fixing stuff" like televisions and stereos. He would make repairs for his friends and neighbours; he would also take apart the instruments in his father's chemistry lab, being careful to reassemble them.

In 1978, at the age of sixteen, Ross received from one of his dad's college friends a set of about fifty logic chips, which are used to program or provide instructions to a computer. By age seventeen, he had designed and built numerous logic circuits that ranged from simple arithmetic logic units to complex microprocessors. It was then that he knew engineering was for him.

In 1980, Ross attended Carleton University and his favourite professors were those who connected the world outside the university to the classroom. With their guidance, Ross organized an event for students to meet professionals from industry who would tell them about their jobs and offer practical advice. At the same time, he got a job at Intellitech Canada Limited, where he gained experience working on audio signal processing and communications with Kal Toth, who taught at Carleton University and was the founder of the company, a systems engineering and consulting firm.

Another mentor, Spruce Riordon, dean of the school of engineering at Carleton University, taught Ross that engineering was not just about technology. He also emphasized the importance of creating a safe environment, of valuing people and demonstrating to them that their concerns had been heard. This demonstration of leadership made a lasting impression.

When Ross told his parents that he wanted to go to graduate school, they were delighted, although he jokes that they were probably just relieved because, even as a child, he had a mind of his own. In 1984, he was accepted at Stanford University, where he studied aerospace. His advisor was Dr. Robert Cannon, professor of aeronautics and astronautics, who was known for his visionary work on robotics and for having proved Einstein's theory of relativity. He had also worked in government, including as assistant secretary of transportation for R&D under Richard Nixon. Cannon's philosophy was one that Ross earnestly shared: theory can be useful, but actually demonstrating its validity is important. You must be able to build a robot *and* do math. He believed that students could learn much from the entrepreneurial world, and he wanted to create the next generation of commercial leaders.

At that time, Stanford's Aeronautics and Astronautics Department admitted students for master's degrees, after which only a few would be considered for admission to the PhD program. On completing their master's degrees, students would spend one and a half to two years studying for their qualifying exams. Each student would be interviewed by a series of six professors, each for one hour. Ross took the exam with eleven other students and only two passed. Ross was one of them.

In his master's year, Ross and a colleague were frustrated that IBM PC XT computers were not as fast as they could be. He realized that the problem was with the hard timing signals: to save money, IBM had tied together the clock signal for the microprocessor and the DMA controller. The slow speed of the DMA controller held back the CPU. Ross and his colleague co-founded SpeedQuest to make and sell a small plug-in clock switching board that could move smoothly between two different frequencies, not integral multiples of each other. With it, computers could run 40 to 60 percent faster. IBM's next-generation computer eliminated the problem, but until then, Ross's innovative "fix" was in high demand.

While preparing for the PhD qualifying exams, Ross was approached in the 1980s by Ned Lerner, a video game programmer who established a 3D simulation development company. Lerner wanted to make the next big flight simulator game: Chuck Yeager's Advanced Flight Trainer. Ned had reached out to folks in the computer science school, but to write this kind of code, you needed to understand aerodynamics. Then he identified Ross, who created a parametric flight simulator that could mimic the behaviour of biplanes, jet planes, supersonic fighters, and the space shuttle. It needed only 2 percent of a PC computer to run. To do this, Ross reviewed the fundamentals of all areas of aerospace engineering without realizing that he was actually doing the best possible preparation for his exams. The game sold over a million copies. Ross ended up graduating from his doctoral program with money in the bank.

In 1990, Ross set out to find a job in aerospace engineering. However, the Cold War had ended and there were no jobs available in the field. So instead, Ross co-founded CriSys Limited in Toronto to bring AI to the 911 dispatch processes, which were then designed to create good, legal records of calls, but not to help the people in crisis on the line. By innovating the system and the process, CriSys's fire department customers reported improved overall response times by about one minute, which can be the difference between life and death.

While in Toronto, Ross met his future wife, and after several years they made the difficult decision to move the family to California. Ross hated to leave his colleagues and business partners, but California was where important innovations were happening, and he really missed the climate. In 1997, Ross, a dual citizen, packed his car and drove to California to look for employment opportunities. A few months later, he began to work for On Your Mind, a company started by friends. His task was to design a program that would replace the customer-client interaction in specific decision analysis scenarios. Ross quickly understood that the value proposition of the company, what customers truly appreciated, was human interface. He spoke with the consultants and told the founders that the web interface would likely not achieve their goals of scalability.

In 1999, he moved on to BlueDog, a dot-com start-up where everyone was under the age of twenty-five. Employees worked Monday to Friday, 10 a.m. to 9 p.m., and it was more or less understood that "if you did not work on Saturdays, you should not bother coming in on Sunday." The company failed to raise sufficient funds in the dot-com crash, so in 2000 Ross moved on to Google.

Ross was Google's first director of engineering, and he believes the second oldest employee at the time. He was not trained in software infrastructure, but in multidisciplinary aerospace and robotics engineering. He soon proved himself useful as one of the early inventors of AdWords, the main revenue driver of the company. This helped Google grow its own services and data centers. The company needed to hire more people and often included Ross in the interview committees. As a result, his expertise in multiple areas began to be noticed, and he was drawn into a variety of projects, including one on renewable energy. The project focused uniquely on developing solar technology, but Ross realized that solar technology alone would not be sufficient to solve the energy needs of the world's population. In 2014, he co-wrote a seminal article titled "What It Would Really Take to Reverse Climate Change." He thought about nuclear fusion and the timing was perfect. In late 2014, he was given funding and the mandate to lead Google's nuclear energy R&D group. It collaborated with TAE Technologies, helped start a policy initiative in Washington, DC, for advanced nuclear research, and funded a scientific investigation into the circumstances under which cold fusion claims were made. Hot fusion occurs under great heat and pressure, and the energy is similar to that of exploding stars in the universe or hydrogen bombs. It was reported that a lab had produced fusion at room temperature. This claim was never replicated and generally considered incorrect. Nevertheless, it was deemed worth verifying. Following the investigation, scientists now concentrate on hot fusion. Google itself is investing in two fusion companies. Nuclear fusion has the potential to produce an enormous amount of energy, enough to solve all the world's energy problems. It uses hydrogen isotopes found in nature. The level of radioactivity is low, producing less waste and taking a shorter time to decay. In addition, it would be less expensive than fission. Fission is the process currently used in nuclear power plants and, after a few accidents such as

Chernobyl and the Fukushima Daiichi plant damaged by the typhoon and tsunami in Okuma, Japan, when radiation was released, is the subject of anxiety. Fusion is still under development and while the first successes in producing energy have been reported, they are not of sufficient scale to be applied. The research is expensive but promising.

Ross is a champion of disruptive technology that can change industries and possibly tackle climate change. He is a deep, philosophical yet eminently practical thinker, applying fundamental understanding to development through a combination of classical theory of math and engineering, along with practical applications. He is not simply a thinker; he is a doer, even installing clever energy systems in his own home, winning his city's Green Building Award.

Ross has a long list of publications and patents, an impressive record of success, and the energy and wisdom to consider any failures as fertile grounds for successful innovation on a grander scale. He has even directed and produced films to educate the public on the crisis we face with the environment and the potential of developing energy sources. He attributes his success to a solid cross-engineering background that made it possible to be involved with computer games, aerospace research, 911 dispatch services, Silicon Valley start-ups, and clean energy. If we wonder about his resilience and his humanitarian concern for the world, we need only read the eulogy he wrote for his mother when she passed away. He reminds us that she grew up in a time when there were very few opportunities for women, yet she used her linguistic talents, speaking four languages, to secure work as a translator across multiple sectors. Seeing the need in her community, she volunteered to help resettle immigrants in Canada. Ross may well have inherited his mother's ability to translate his talents across fields and her desire to serve humanity.

25

ANNETTE VERSCHUREN

Leading Corporate Social Responsibility

Cape Breton, NS

Chair of the MaRS Discovery District board, founder and president of NRStor Inc., former CEO of Home Depot Canada and Asia, former president and co-owner of Michaels Canada, member of many business boards, recipient of over a dozen honourary degrees, author, chancellor of Cape Breton University, Annette Verschuren likes to remind people that she grew up on a farm in Cape Breton, where she learned to be innovative, assertive, and to care for the environment.

A life-changing moment came when Annette was eleven years old. In 1967, her father had a heart attack, leaving Annette (the third of five children), her siblings, and her mother responsible for running the farm. Before the heart attack, her father had provided them lessons in innovation. In 1964, long before it became popular, he installed a stereo system in the barn, providing relaxing music for the dairy cows and increasing production of milk. As a result, Annette and her siblings did not hesitate to try new ways of accomplishing the tasks before them. They found, for example, ways to hoist items that were too heavy for them to lift. Realizing that she might make extra money babysitting, but knowing she had to do the farm chores, Annette brought the children to the farm and watched over them while she worked. In addition, Annette's mother kept the books for the farm and shared her good financial sense with her daughter.

Annette was an excellent student, and in primary school she became the teacher's helper, working with students in one room of the two-room schoolhouse while the teacher gave lessons in the other. Annette loved having the responsibility and the opportunity to play a leadership role. At the age of sixteen, she had two major kidney operations, which forced her to miss nearly a year of high school. One of her teachers, Gwen Sheperd, who was also a busy mother with two sons, invited Annette to her home every Saturday morning and spent two hours helping her with math so she could catch up on what she had missed.

When Annette was planning to attend university, there were three career paths open to women: teaching, nursing, and secretarial arts. She thought she might become a teacher like her sister but, on entering St. Francis Xavier University in Antigonish, Nova Scotia, she switched her major to business. She neglected to mention this to her parents and told them only over the Christmas break. They encouraged Annette to continue doing well and she did—until she got a less-than-satisfactory grade in auditing. The professor called her in and said, "You'll never make it. I will help you fill out an application for secretarial arts." This comment did not discourage Annette; if anything, such barriers helped make her the person she has become.

Starting at age sixteen, Annette worked her way through school and university at CN Marine, where her job was to put railcars on ships, serving night shifts under the master foreman. After graduating with a business degree, as one of three women in the class of 1977, she went to work as development officer for the Cape Breton Development Corporation, a coal-mining company. Miners were superstitious and believed that women should not go down into the mines. Annette nonetheless descended into the mines to execute her duties and demonstrate leadership. The miners expected to see their bosses, but did not expect to see her. She wore a kerchief under her hard hat and did her best to make everyone feel at ease, paving the way for other women to be accepted in the mines to work and occupy leadership positions.

In 1986, she worked for the Canadian Development Investment Corporation as executive vice president privatizing Crown corporations. She followed this job in 1989 by serving as vice president of development for Imasco, a major corporation that owned Canada Trust, Imperial Tobacco, Shoppers Drug Mart, Genstar, and the Hardee's restaurant chain. In 1993, she then brought the arts and crafts chain store Michaels to Canada as president and co-owner. The next challenge she took up was serving as CEO of Home Depot for Canada and Asia, in 1996 moving the company from nineteen to one hundred seventy-nine stores.

Throughout her career, Purdy Crawford was more than a mentor. A lawyer born in Nova Scotia, he was president and CEO and chair of the board of Imasco. He encouraged her to leave Cape Breton for Toronto to pursue employment opportunities. He was ahead of his time in recognizing women's potential and in pointing out pathways to success.

Over the years, she has had the occasion to work in China and has visited sixty-two countries. Her observations of best practices and ideas from around the world have served her well, teaching her many things that might be adapted and adopted in Canada.

Annette's concern for the world and the environment is palpable. She believes we need to reduce carbon consumption, to move away from our consumer habits, and to electrify buildings and transportation. She hopes that "it will not take a disaster to bring back the collective will of the people of the world." She also hopes that citizens around the globe will become more responsible for one another and for the planet.

In every one of her positions, she has worked to effect positive change in sustainability. For example, when she was CEO of Home Depot, the company joined the Forest Stewardship Council and stopped clear-cutting trees. NRStor Inc., the clean energy company she founded in 2012, has big projects for energy storage and is working with the Inuit population to provide sustainable energy. The town of Arviat, Nunavut, located on the shore of Hudson Bay, will be a first adopter using a combination of solar, wind, and battery power, creating a model, sustainable community. The company is also working with the Six Nations of Grand River on a major installation that will store clean energy, saving from $700 million to $1 billion for taxpayers. The company also adopted the first commercial grid, connecting an energy-storage flywheel in Minto, Ontario.

Annette believes that greater diversity in the workforce and in leadership roles will bring about positive change. She ventures to say that if half of the world's leaders were women, the outcomes would be positive and most welcome. She thinks that we need to listen to others, even if they have conflicting opinions. Annette says that it is "important to get along with others and to adjust to their personalities and understand their cultures. We need to integrate people from different fields to effectively find solutions to the problems that beset us." She herself has often changed her perspective after listening to others.

When asked for advice, Annette muses, "If I could do it all over again. . . . Well, I would still go toward the challenge of the future. Today that is climate change, and it will remain a challenge for the human race for many years. I would prepare myself to help by studying chemistry, physics, mathematics, and emerging technologies. Work will always be important so I would advise young scholars to take advantage of a co-op program. Travel is a good way to learn and gain a broader perspective. Information technology and cybersecurity will grow in importance."

While she worries about the future, Annette is hopeful. She recognizes that humanity has the potential and the ability to change and to contribute to a better, more positive future. Annette personifies that ability and that potential. Her dedication to the environment and her leadership are her gifts to our world. Her many contributions, from saving forests to making it possible for communities to adopt clean energy, stand as innovative examples of the change we must adopt.

26

ROBERTA BONDAR

Making Space for Women

Sault Ste. Marie/Toronto, ON

When Roberta Bondar was a child, growing up in Sault Ste. Marie, Ontario, in the 1950s, she loved to go camping because she could look up at the stars. She also used to visit her neighbour's house to watch black-and-white Flash Gordon movies on television. With her sister, she repurposed a large wooden shipping container that her father brought home from his office and painted one side to look like a spaceship and the other to resemble a submarine complete with a periscope, and then mounted it on sturdy wagon wheels. Creativity was valued in her family. Her father further encouraged her by building a small science lab in her grandparents' basement.

When not playing sports, she liked being outdoors, and her parents gave her Rachel Carson's *Silent Spring* and a book on the stars and planets. They also bought both science books and encyclopedias. *Silent Spring* and the little book on stars still sit on her bookshelf, reminders of the love and support of her family and her inspiration and determination to persevere.

Although she liked physics, math, and physical education in high school, she found that her English and French teachers were the most influential. They enthusiastically introduced her to literature, travel, and history, and taught her the importance of words and communication. Despite their best efforts, Roberta still preferred science. There was a life-changing moment when her mother stood up to the school's vice principal and guidance counsellor to insist that her daughter be allowed to take math and science in grade thirteen. In particular, Roberta had a passion for flight, stars, the constellations, and the universe. In the 1960s, there was no opportunity in Canada for a female to serve in the Royal Canadian Air Force or to pursue a career as an astronaut. In fact, women were not visibly present in space-related professions anywhere. So, while she did not forget her dream, she went to the University of Guelph to study entomology, a field of interest because of her work as a summer student at a local federal government insect laboratory in Sault Ste. Marie. While at university, she became ill with a case of mumps and when she was released from the infirmary, she got a room in the home of a mature student who allowed Roberta to use her microscope and other equipment. Roberta became fascinated with embryology, the study of the embryo and fetus, and histology, the microscopic anatomy of the tissues, and changed her major to zoology. In 1968, she completed a double degree in agriculture and science, specializing in zoology. Her embryology professor encouraged her to go to the University of Western Ontario in London, Ontario, for a master's degree in science, and in 1971 she completed a degree in experimental pathology. In 1974, she graduated from the University of Toronto with a PhD in neuroscience, and in 1977 she completed an MD degree at McMaster University in Hamilton, Ontario. She then did postdoctoral fellowships in neuroscience at the Toronto Western Hospital and in neuro-ophthalmology at Tufts New England Center (now Tufts Medical Center) in Boston. Her career path in clinical medicine and medical research was laid out in front of her, but still never far from her mind was the dream of becoming an astronaut. Then, in 1978, six women astronauts were named by NASA: Sally Ride, Shannon Lucid, Judith Resnik, Anna Lee Fisher, Margaret Rhea Seddon, and Kathy Sullivan. They were selected to fly on the Space Shuttle. As she watched the televised first landing of the Space Shuttle, Roberta told herself, *I am going to do this!*

She did not have the necessary funding to go to the United States. So while in medical school and during her residencies in Toronto and Boston, she limited her expenses and continued her studies despite considerable financial hardship. She wanted to be an astronaut, but had no idea of how to make that happen. However, while doing her postdoctoral fellowships, she obtained a license as a private fixed-wing pilot, and flew a plane whenever she had saved a few dollars to pay the fees. Then the call for astronauts in Canada in 1983 changed everything. From 4,300 applicants, of whom 10 percent were women, they chose Roberta as the only woman and one of the six original Canadian astronauts.

The Canadian Space Agency wanted the astronauts to wear a uniform to events. While the male astronauts wore pants, shirts, and ties, Roberta was given a skirt and a scarf to be tied into a bow. During the press tour, she felt that no one took her seriously and vowed never to wear a skirt again. And she never has.

In fact, she set a bit of a trend. Before her mission, the astronauts were asked to take an informal photo for which the six members of the crew—five men and Roberta—could select an outfit that best represented them culturally or historically. The German astronaut from the European Space Agency came in traditional lederhosen, and Roberta dressed as an RCMP officer, but instead of wearing the high heels and the skirt that female officers had to don for formal dress, she wore breeches and the Red Serge. The picture became quite well known, and Roberta found herself a poster person for the RCMP's female officers, who wanted the right to wear the same dress uniform as their male counterparts.

Training for the mission included work with the European Space Agency, as she performed experiments for them while in space. These experiments included studying the mental and physical effects of the lack of gravity and ways to treat injuries in space. Besides working on the experiments in space, Roberta had time to peer out the portals and to see the Earth from a different perspective. She recalls looking down on the Earth and reflecting on how tiny it seemed. She recognized features like the Aswan Dam, but also thought about how distance blotted out features like poverty as well as the people working to alleviate it, trying to make a difference. She thought about birds flying thanks to their feathers and hollow bones, and about tremendously large machines like airplanes that are able to fly thanks to the ingenuity of humankind. She observed weather patterns that affected both the environment and people, and she meditated on the importance of contributing positively to the world. She felt incredibly small, but thought we should constantly work to create a better environment and look inside ourselves and to the past of civilization to find inspiration.

On her return to Earth, Roberta decided to help people understand, in the best way that she could, the incredible beauty of our planet and the importance of the natural world for good mental health and for stimulating curiosity and innovation. She became a photographer and circled the globe, creating exquisite photos in places where people rarely travel, which were made into an exhibit and a book. Then she dedicated time and resources to teaching young people around the world to respect and appreciate nature; she hopes that they will protect the environment in the future. She flew with young whooping cranes, an endangered species, leading them south, teaching them to migrate in the absence of their parents. Roberta shares with us the beauty and the fragility of the world as she sees and has seen it from a unique perspective.

Space helped Roberta think about the meaning of life and human civilization. She believes that we need "to look inside ourselves and create ourselves without losing ourselves." She says that we should endeavour to create a future of which we can be proud. And we must remember that time is a gift we must use wisely. Roberta employs not only her time but also her wisdom and skills to opening our eyes to the beauty of our planet in need of preservation.

27

GEORGE ZHU

The Sky Is No Limit

Toronto, ON

Born in Shanghai in 1961, George Zhu came from a family that supported the belief that education offered the best possibilities for the future. They encouraged him to study, but he was born just before the Cultural Revolution, which favoured socialism and Communism and opposed capitalism and democracy. Intellectuals were imprisoned or sent to work on farms, and universities were shut down by Communist leader Mao Zedong. George's parents' hopes for his future were dashed.

In 1969, some elderly neighbours told George about Apollo 11, the US's first successful spaceflight that landed humans on the moon. Thereafter, George would look up at the moon and stars and dream about space travel.

In 1978, schools in China reopened and George, then seventeen years old, redoubled his efforts to learn and pass the national entrance exam. In 1979, he was admitted to Shanghai Jiao Tong University, where he began as an undergraduate and graduated with a PhD in 1989. He was not able to study aerospace at that time because China was focused on economic development. So he studied engineering mechanics and worked for several years in a company before being invited to the University of Toronto in 1993 as a research associate, where he studied lighter and stronger materials for aerospace engineering and subsequently worked at Indal Technologies

Inc. (now Curtiss-Wright Indal Technologies, an aerospace company) in Mississauga, Ontario, where it built secure and traverse systems for assisting helicopter landings on ships and cable-towed sonar systems.

While he was working, he completed a second master's degree in robotics at the University of Waterloo in 1995, and another PhD at the University of Toronto, where he studied aerial refueling and proposed a solution to the problems that had been experienced with aerial refueling hose systems.

Recruited as a professor of engineering design at York University in Toronto in 2006, he started a new career in astronautics and became the director of the space engineering undergraduate program. His innovative thinking led him to propose using the tether dynamics employed for refueling airplanes and towing sonars in the sea for space tether systems, in particular, electrodynamic tethers, which are long, electrically conducting cables operating on electromagnetic principles to convert their kinetic energy to electrical energy or vice versa when they pass through a planet's magnetic field. When converting their kinetic energy to electrical energy, the electrodynamic tethers lose altitude, as they would if experiencing a drag. George thinks this function could be used to remove space debris, such as dysfunctional satellites and spent rockets that currently endanger operational satellites and the International Space Station. George's method will also achieve deorbit without using propellant fuel—the electrodynamic tether system can catch debris and accelerate its descent into the atmosphere. In 2020, George was sponsored by the Canadian Space Agency to try this technique using two CubeSats, which are miniature satellites weighing less than two kilograms that are used for communications and remote sensing. However, the tether did not deploy due to a technical failure. George is certain this technology will work and eagerly awaits the next opportunity to test it.

George gets some of his best ideas while listening to others. For example, in 2012 he attended a conference in California on space colonization. There were delegations from various trades, including carpenters, proposing that they and their materials be taken to Mars. As he was listening, George thought, *Why do we need to carry everything to Mars?* A rocket has limited volume. Whatever is taken to space must be very small, or foldable. If equipment is deployed for a single use, it would be wasteful. We currently print in 3D on Earth, and we may well be able to print in space one day.

George then asked, *Why not bring 3D printers to space along with some raw material?* Astronauts could print parts that might have broken, and they could manufacture almost anything they needed. George then imagined a 3D printer that would work in zero gravity and in a vacuum. Obviously NASA had a similar idea because in 2015 it sent a 3D printer into space, but it did not work. George is collaborating with colleagues at York and Western Universities on a 3D printer that will function in zero-gravity conditions and will print biological tissues that can be used to heal astronauts should they get injured while on a long mission. The printer's bio-ink will contain living cells and be able to generate bio-tissues. On Earth, softer tissue requires a scaffold or a structure, but this is not needed when there is no gravity. George is most proud of this collaboration and believes that in the next decade we will see 3D printers being effectively used in space.

Over the next ten years, George wants to participate in the Lunar Gateway program by developing new robotic technologies using artificial intelligence. He foresees a workstation orbiting the moon and imagines that one day people will go there to establish life. He also predicts that in another twenty or thirty years there will actually be an effort to establish a base on Mars.

His advice to future aerospace engineers and, for that matter, to all young students is "If you have a dream, insist. Do not be afraid of other people's comments." When the concept of space elevators was first mentioned, people laughed and said it was just "sci-fi." But George and his students are working on an innovative space elevator now. It will use a tether as a kind of springboard and will not at all resemble the elevators on Earth. He has received calls from investors in California who want to know more about the space elevator. He says, "Be persistent and keep working on your concept. If you do not believe in yourself, how will you convince others to support you?"

George is thinking all the time. His ideas are not the result of a miracle, but of a lifetime of thought and experience. When he needs inspiration, he sits down and talks with his graduate students. As their professor he often wonders whether he is actually teaching or learning from his students. He also finds that taking a long walk helps ideas to coalesce. "You cannot plan for innovation," he says. "You plan for 'A' and sometimes you end up with an innovative idea." George says that space is the new frontier and there are thousands of possibilities. Just think, he says, "thirty or forty years ago there was no GPS. It is hard to imagine travelling without a GPS nowadays. It should not, perhaps, be so difficult to imagine space travel thirty or forty years from now. Today we are entering a new era of electric vehicles. Battery technology and self-driving vehicles are avenues that will be developed. If we can change the entire automotive industry, surely we can create a few labs in space."

28

MIKE PALMER, CHRIS BURN, STEVEN KOKELJ, AND HEATHER JAMIESON

The Mentors

Yellowknife, NWT; Ottawa, ON; Yellowknife, NWT; Kingston, ON

The paths of four innovative researchers—experts in northern geology, permafrost, climate change, mining, and pollution—are intertwined not only through their care for the environment and the northern communities where they live and work, but through mentorship and friendship.

Manager of the North Slave Research Centre for Aurora College, Yellowknife, Northwest Territories, **Mike Palmer** was born in 1977 and grew up in Ottawa, Ontario, in a family that encouraged him to study and embrace outdoor activities. His most memorable childhood experience is a canoe trip down the white water of the Mountain River in the Mackenzie Mountains, which he took with his father when he was sixteen years old. The thousand-foot-high peaks rising from the shore inspired his lifelong love of the northern landscape. When he was a teenager, Mike worked at a summer camp, and before earning a BSc in earth science and Spanish at Dalhousie University in 1997, he took a year off to travel to India and Nepal, inspired by articles he read in *National Geographic*. After university, he worked as an instructor for Outward Bound, travelling and kayaking across North and South America. In 2004 he met Professor Chris Burn at Carleton University and decided to pursue graduate studies, "running the academic gauntlet" under Chris's guidance and completing his PhD in 2022.

Chris Burn is *the* geology professor par excellence and *the* mentor we all hope to have at some point in our lives. He is an outstanding researcher and scholar, and his dedication to his students and the principles he instills in them are truly remarkable.

Born in England in 1959, Chris grew up in a small town eighty kilometres from Cape Comorin (Kanyakumari) in India. In 1971, his family returned to England, where he earned his first degree in geology at Durham University. Inspired by his father, a professor of mathematics and an Anglican priest, and his mother, who studied classics and anthropology and was engaged in social work, Chris learned the importance of connecting with and respecting people. He recalls two special teachers. He met the first in elementary school, Margaret Somerville, and she taught him to read, not by recognizing words, but by giving him books he would enjoy reading. They remained in contact over many years. The second, Tony Billinghurst, was a French high school teacher who believed in outdoor experience and took his students to the Cairngorms, a mountain range in the eastern Highlands, Scotland, to hike and learn survival techniques, even in winter. He gave Chris a love of the outdoors and provided a model of caring, generously supporting his students with railway tickets and tents.

Chris came to Canada in 1981 to attend Carleton University on a Commonwealth Scholarship. After completing his PhD in geology in 1986, he did postdoctoral research on permafrost environments at the University of British Columbia in Vancouver with Dr. Ross Mackay, the world's leading authority on permafrost, who became his mentor and collaborator. Their last joint publication appeared in 2013, when Ross was ninety-seven.

Chris invited his student Mike Palmer to participate in extended fieldwork in the Inuvialuit Settlement Region on Richards and Garry Islands in the outer Mackenzie Delta. He suggested that Mike take on an internship with Steve Kokelj, one of Chris's students who had recently received his PhD. They might work together to look at permafrost conditions along the proposed Mackenzie Gas Project corridor. The internship morphed into a job and Mike never looked back.

Steve Kokelj was born in a rural environment north of Toronto, Ontario, in 1972, and his family roots are in Sudbury, Ontario. As a youth, his passions were hockey and the outdoors. One of his best memories is exploring the Credit River in a canoe. He earned his BSc at Carleton University in 1996, his master's in geography at the University of Ottawa in 1998, and his PhD in physical geography at Carleton University in 2003. He discovered his true passion for geography and geology while doing fieldwork in the Yukon, collecting and analyzing data on permafrost. He recalls being sent to Ellesmere Island for three months in 1996 to collect data with another student, a brilliant hydrologist. They had been friends for many years and were both completing master's degree projects. She then took a job in Yellowknife and Steve followed her. They married in 1999 and have lived and worked in Yellowknife ever since.

Before they got married, Steve first had to complete his comprehensive exams with Chris Burn, whose directed reading course meant studying five papers every week, carefully selected to reflect the student's interests. The papers had to be studied meticulously in preparation for a two-hour talk with Chris, followed by a three-hour exam. Steve can still quote from the readings today.

A few years passed before Steve asked Chris why he posed the hardest questions of all the examiners at his thesis defence. Chris responded with a question: "What would a good coach do? Would he push you or not? I wanted to see how far I could push you because I wanted you to continue researching and thinking in the future."

In 2012, while working on the decontamination of northern lakes (and there are some 1,800 lakes within thirty kilometres of Yellowknife), Mike cold-called Heather Jamieson, a professor in the Department of Geological Sciences and Geological Engineering and the School of Environmental Studies at Queen's University, Kingston, Ontario, who was working on the arsenic that leached from former mines into the water and soil. They shared research interests, and Heather was the reason that Mike returned to Ottawa to complete his PhD in 2022. Her encouragement, support, and advice were important to him. Heather says it was obvious that he should complete his degree. After all, he had already published about thirteen papers.

Heather Jamieson was born in 1953 and grew up in Rouyn-Noranda, a copper-mining town located in Northwestern Québec. She loved science, and in 1970 she got a summer job mineral exploring for Noranda Mines Ltd. When the mining company hired its first woman geologist, they decided she needed a female assistant and Heather was recruited, along with a third woman, a PhD student from France. The three would sit on a rocky outcropping in the "bush," as it was called, and talk science. They also spoke of the hardships faced by women in a field like geology, where less than 10 percent of the employees were women. These women became Heather's first mentors.

She went to the University of Toronto and received her BSc in 1976. She was interested in mineral exploration and thought of travelling to interesting places. At this time, environmental geology was beginning to be recognized as an important field of study. Heather's professors suggested that this field might be good for "girls" because it was a "soft science" and did not require travel to the "boonies." Heather did not want to do "soft science," and chose the area the farthest from the suggested field for her doctoral studies: geochemical thermodynamics, receiving her PhD in 1982.

However, she became passionate about the environmental impacts of mining and changed course, heading to Yellowknife in 1999 to study the impact of the arsenic that had been released into the environment by gold mining. She recalls flying to the Yukon and hearing from a fellow passenger about a legendary professor at Queen's University who had discovered the gold deposits around Yellowknife. The passenger said that some community members wished he had never found them because of the subsequent damage the mining companies inflicted on the environment. During the active gold mining years and even later, there were no environmental controls, nor was it understood how the arsenic was released. In 1999, Heather initiated a study of Giant Mine, aptly named, as it was one of the largest. She consulted with members of the Dene First Nations community, who described how they no longer picked berries or drank the water in certain areas because they made people sick.

Mike's research suggested that the arsenic spread much farther than was previously thought. Heather was able to prove that the arsenic permeated the water and soil up to thirty kilometres from a mine site. Steve is now looking at the contaminants in the permafrost that are melting into the rivers and lakes. Mike has overseen, with graduate students and community volunteers, the planting of over twenty community gardens in areas where the soil has not been disturbed, testing the produce they grew for arsenic. As manager of the North Slave Research Centre, he oversees STEM outreach in twenty-five locations, including a number of Indigenous communities, with which there are also a number of collaborative research projects.

Heather is grateful to her mentors, who include Ursula Franklin, an internationally known metallurgist and physicist based at the University of Toronto, whose interests included the political and social effects of technology on the environment and the population. Another mentor was Barbara Sherriff, professor at the University of Manitoba and environmental consultant who specialized in the delineation of acidity and salinity of gold mine tailings. Heather notes that she herself has been part colleague, part employer, part supervisor, and part parent to many students. A recipient of many grants from the Natural Sciences and Engineering Research Council (NSERC) and the Canada Foundation for Innovation (CFI), Heather is most proud of receiving the Mentorship Award from Women in Mining Canada.

Heather recently retired, and the last course she offered was an interdisciplinary environmental exploration of the last five extinctions on Earth, the most recent being dinosaurs. The sixth extinction would be humankind. Former extinctions were caused by external forces like volcanic eruptions or asteroids hitting the Earth. The final extinction will likely be caused by humans themselves—unless they can find solutions. And she works with the students she has mentored to do just that.

Steve and Mike remain in close touch with Chris. Steve studies permafrost at work and at home (his house is built on permafrost). He currently mentors five students from across Canada and Mike has joined the group. Steve says we need to think of permafrost as a nonpermanent component of the environment, which is influenced by global warming. There have been increasing numbers of landslides around small lakes, caused by the melting permafrost. When the Laurentide ice sheet retreated, hundreds of square kilometres of transformed ice were left and preserved in the permafrost. When the ice melts, landslides and foss slumps (the resulting debris and troughs) will have increased a hundredfold across Western Canada. Steve says the situation is urgent and believes that mentorship will help create an ecosystem of researchers who will work together with the people who live there, so that science in the North can be northern-based science.

"For the North to transform," he says, "it has to get through its colonial past and achieve reconciliation by rediscovering and building on its capacity and this can definitely occur through mentorship."

While academics know best how to describe the northern environment as it was one or two hundred years ago, he notes that the best hunters in the local community provide the most accurate observations on the environment now. They use multivariate statistics, the variations in observed data in their heads, and can tell you how environmental conditions have changed.

Chris teaches his students about investing in a place and the people. The environment is not disconnected from those who call it home. Mike agrees.

"Science is not just about writing scientific papers and reports," he says. "I would like to see us develop a culture of science that recognizes people's connection to the land and to their history. We need to build a scientific culture within existing institutions. This requires bold vision and leadership. People are the key. We need highly qualified people, and we must ensure that those living in our communities have the opportunity to take on these roles."

Mike says he owes his success to his generous and inspiring mentors and tries, in turn, to teach young researchers. He has built a career as a scientist rooted in his community. He may not have experienced an aha moment, but he enjoys an impressive international reputation, and his patient dedication to his research and his students contribute to a creative and inclusive environment for learning.

Heather did groundbreaking work on the way arsenic permeated the water and soil, involved Indigenous residents of the region in her research, chose "hard science," and did fieldwork at a time when women were not welcomed. She mentored students before mentoring became popular. Her life is a story of innovation, strength of character, and passion for science and the environment.

Chris says, "The North is a home. People live there. It is their place. We are here in the university to serve the people of Canada and that means the

North. It is an extraordinary privilege to work with people in the North and one must approach them with humility." Chris is an extraordinary mentor and teacher, and it is no surprise that Mike and Steve consider him a member of their families. Thanks to Chris, Heather, Steve, and Mike the next generation of researchers will be known for their discipline and compassion. The North and the entire country will benefit.

PART FOUR

Quantum Basics

Quantum theory, computing, and mechanics encompass the search to understand the universe, the race to create the perfect quantum computer, and the application of quantum mechanics to the transformation of entire fields of human activity, including banking, health care, communications, and security.

Now, before your eyes glaze over, please take a sip of coffee or your favourite beverage and continue reading. I promise to make this light and will begin by renouncing the use of words I had to look up myself!

In the eighteenth century, the philosopher Voltaire imagined humankind to be located somewhere between the infinitely large and the infinitesimally small. Both held the answers to many questions, but they remained mysteries, as we had not yet developed the science that permitted us to see further into the universe or more of the tiny particles that make up the Earth.

Today we have identified planets outside our solar system, and we still wonder what they might really be like. We have also discovered even smaller particles than the ones I learned about in school. Tinier than atoms, neutrons, protons, or electrons, these particles do not follow the neat attraction/repulsion, positive/negative actions I studied many years ago. Indeed, they keep strange company, travelling alone, in pairs or groups. When they once connect, they will identify with the other particle, no matter how distant in time or space. The separated particles will respond in like fashion to a stimulus. It is like speed dating in that it leaves a lasting impression that is so strong that every time the participants hear the theme song, they will perform the dance moves whether they are in a disco or an elevator thousands of kilometres apart.

The discovery of these new particles led researchers to attempt to harness them with the goal of improving computing power. In order to develop the quantum computer, a perfectly controlled environment is needed, usually a vacuum in a location as free of vibrations as possible so that the reactions of the small particles can be limited and consistent. This is, in a few words, the race for the quantum computer. Can the tiny particles be harnessed? Can the size of the machine be reduced or expanded? Can scientists figure out how to capture the power of these particles? How can this be used in the future? With which materials?

I should add that the situation, even inside the vacuum, is messy. If we imagine dropping a pebble in a pond, we see the splash, the waves in expanding circles going to the edge of the pond and turning back on themselves. This is the horizontal dimension but vertically, as the stone descends to the bottom, each level produces a new wave under the surface, which then moves out and back, following a schedule slightly different from the circle we see on the surface. When the stone touches bottom, a new shape of competing waves rises toward the surface. When researchers direct a beam through the vacuum, it causes the very small particles to move. The atmosphere of the vacuum is controlled but still imperfect, and as a result, the movement of the tiny particles is not always predictable.

Why would one want to create a quantum computer? Because one can solve larger, more complex problems faster than ever. A quantum computer will be able to deal with more data than ever could be accomplished with existing machines. They will do so with incredible speed. The problem at present is that researchers have not yet completely understood or controlled the movement of these particles and, combined with the impossibility of creating the perfect state of motionlessness, the results are sometimes brilliant and other times incorrect. However, the great race is on. It is a bit like the old film *Around the World in 80 Days*. In labs around the world, scientists are getting closer every day to *the* quantum computer. (The quantum computers that already exist are not yet *the* holy grail, but they come very close and are most impressive.) I have little doubt that one of these days the good news will be on the front page of every newspaper in the world.

The technologies that have been unleashed with the power already existing will explode and become truly extraordinary when *the* quantum computer is created. Researchers will likely be able to simulate and design specific drugs, produce "green" hydrogen, or design more effective agricultural processes. Environmentalists will be able to collect and analyze data in real time so that we can learn how to survive climate crises and to prevent them in the future. As with all new developments, quantum technologies will also be fraught with dangers and possibilities. For example, if quantum is able to break the security keys of computers, the world's data will be at risk. On the other hand, quantum could (and some say definitely will) lead to a more secure way to protect data based on quantum mechanics.

Quantum theory has enabled astrophysicists to verify other existing theories and to gain insights into black holes. Quantum computing is already inspiring researchers to imagine what they can learn from massive amounts of data to make our lives better. Quantum mechanics and communications will enable us to harness the power of these tiny particles and apply them to new understandings of many fields, including economics and transportation.

Many times in the past, individuals have seen the possibilities and potential of harnessing new sources of energy, while much of the population has remained incredulous. I am thinking of the steamboat that was promoted by Robert Fulton in the US. Back in 1807, it was known as "Fulton's Folly." That was until it sailed up the Hudson River from New York City to Albany in an impressive thirty-two hours.

Canada has been a leader in the design and development of several types of "quantum computer." As was once the case for Fulton, I hope that Canadian researchers will receive well-deserved recognition for the progress they continue to make.

Canadians can be proud of the work ongoing at the Perimeter Institute and universities across the country, including Sherbrooke, Waterloo, Calgary, British Columbia, Simon Fraser, and Toronto, to name a few. The foresight and energy of industry partners at MaRS and Vector are at the forefront of efforts to carry on both theoretical and applied research in this area at the same time. Rather than wonder with disbelief, we must start imagining the possibilities that lie ahead.

29

ROGER MELKO

Finding Algorithms to Accelerate Discovery

Waterloo, ON

Leader of the Perimeter Institute Quantum Intelligence Lab, Canada Research Chair at Waterloo University, chief scientist for Creative Destruction Lab's quantum incubator stream, affiliated with the Vector Institute for Artificial Intelligence, Roger Melko has been recognized with the Young Scientist Award and the Herzberg and Brockhouse medals. His innovative thinking and use of computer simulations have led to the naming of low-temperature spin ice as "Melko State." His research is new, original, and brilliant, unifying AI, machine learning, and many-body physics.

Roger Melko was born in The Pas, a small town 520 kilometres north of Winnipeg at the confluence of the Pasquia and Saskatchewan Rivers, and when I spoke to him, he reminded me, just in case I forgot, that moose hunting season in the North began the day before. His parents were already at their camp, and they might get lucky chasing moose this year. Roger then reminds me that northerners are known for their sense of humour, which probably is developed during the long, cold winters.

During his childhood, Roger fell in love with science at home and in school. In high school he remembers a number of influential teachers who introduced him to science and math. Roger worked summers in boat maintenance, where he learned about the mechanics of engines. When it was time to go to university, he leafed through pamphlets at school and tried to get a feel for what life would be like on a university campus. He chose the University of Waterloo because of its reputation in science and computer science. He loved physics and wondered what he should do. Should he major in physics and minor in math? This problem turned out to be relatively small compared to his culture shock on arriving in Waterloo. It was quite an experience for an eighteen-year-old youth from The Pas to land in Southern Ontario. He thought, *Well, if you can handle this, not knowing what you were getting into, you can do anything.* At university he was part of a co-op program that took him to labs in Ottawa at Health Canada and Carleton University as well as Fort McMurray, where he was able to experience work and to make the tough choice between seeking an academic career and working in industry. He was influenced by his two mentors, who were diametrically opposed in their approaches to science. One concentrated on quantum gravity and quantum mechanics and the other was interested in novel, exotic materials. One was looking at crystal lattices and materials with a high degree of complexity and the other decreased the level of complexity by unifying gravity and quantum theory. Roger was struggling, but his mentors convinced him that if anyone could succeed, he would, and they recommended he do graduate work at the University of California, Santa Barbara.

On arrival in Santa Barbara, he thought the weather and the beaches were great, but he soon found his home-away-from-home in the physics building. He says California was a bigger, faster-moving and -changing world, but he appreciated his education at Waterloo, which gave him the "book smarts" he needed to succeed, just as his mentors had predicted. His PhD was in high-temperature superconductivity theory, and Roger notes that "physics is beautiful" and, with it, he would "like to change the world for the better." Roger says that one individual can have a big influence on humankind. He wants to work on a significant problem that has never previously been explained or designed. He thinks of all the grand challenges we face, and wonders which one, if solved, would best change technology, the scientific landscape, and the world?

On graduation, he was offered a permanent position as staff scientist in computer science and materials technologies in Oak Ridge, Tennessee, where they had just built a new spallation neutron source. He got to see how Big Science is done and he loved his work and the States, but when he was offered a faculty position at Waterloo, he happily accepted and returned to Canada. Roger feels fortunate that his career path made sense—after a bit of soul-searching about the direction. He says he left behind a nine-to-five job for the university life, but feels it actually is not an exclusive choice. Academics can easily collaborate with industry, have their own start-ups, and remain respected scientists.

Past innovations in science, including computers, machine learning, and classical hardware, all have a stake in quantum, which requires an "everyone's in type of approach." Roger feels lucky to have had formal training in condensed matter physics (the complexity approach) and to have been able to cross-pollinate ideas in quantum information sciences. The connection between condensed matter physics/materials and quantum was not previously evident. Eventually, Roger realized that the problem of interacting electrons (in materials/matter) is essentially the same problem as interacting qubits (in a quantum computer) and now is at ease in both fields.

Roger notes that when you introduce new ideas, there are generally two reactions. Some people will say, "This makes no sense and is all wrong," and others will say, "It was obviously right and rather interesting." When you hear people saying both, you know you are onto something! Bringing machine learning and AI into the quantum field was greeted with the standard comments, and so Roger thought that there were strong possibilities for success, and he was certainly correct. He says his work is a process of discovery. He writes problems on the blackboard and then drafts papers for publication. He will sometimes discuss his ideas with young students, who have not yet acquired biases. He likes to work with his team when they find a code is not working or when they see something entirely new.

Roger's goals include continuing to be an educator, helping shape careers, and building a critical mass of innovative people with a diversity of thought. His students are a creative group of brilliant "misfits" who straddle fields and bring expertise in physics, quantum chemistry, statistical mechanics, and computing. He loves research and wants to drive his field forward. He wants his lab to be a greenhouse of ideas in the Canadian winter. He wants to look at ways we can change the world and reduce the number of resource-intensive things we consume and contribute to solving the urgent problems associated with climate change. He loves the fact that new models are always being explored and that Perimeter brings in young people ready to contemplate change in the way we think.

Roger muses, "There are no problems we cannot solve with our minds. After all, problems were created by people. We need to discover that je ne sais quoi." Roger thinks that scientists and investors should ask, When does a quantum computer become advantageous to build? This is not necessarily a question for researchers in the laboratory. It requires microeconomics and game theory. We need to answer rational questions: When is quantum theory valid based on explicit rules? And when is it economically viable to build? Roger wants to find the answers to these questions. He would like to apply a rigorous mathematical discipline to all science projects and use game theory to project the results in five to ten years.

His to-do list includes the development of machine learning algorithms to accelerate discoveries in physics, applicable to building quantum computers and synthetic quantum matter, and the creation of software that will enable the engineering of qubits needed for quantum technology.

Looking into the future, Roger wonders how will a quantum computer emerge from the fundamental equations of quantum mechanics and how does human consciousness emerge from the fundamental biological working of the brain? If physics is the theory of everything and quantum theory is correct, there are still things we cannot explain, including logical rules for superconductivity and AI. Roger has innovated in creating entirely new fields of inquiry and in opening our minds to the possibilities that are almost within our grasp. We will thus be able to harness the potential of quantum to solve problems of climate change and the economy. We most definitely need Roger's brilliance and his optimism to solve the many problems that beset the world.

30

ELIE WOLFE

A Pioneer in Quantum Physics

Waterloo, ON

Working at the intersection of quantum foundations and causal inference (both relatively new fields), Elie Wolfe and his colleagues are pioneers whose work has implications for many other fields, including AI, banking, insurance, climate research, and epidemiology. Elie himself is an expert in quantum physics and studies how quantum resources can generate nonclassical correlations. This means, among other things, that they can adjudicate among difficult-to-discern causal hypotheses involving unobserved factors.

Born auspiciously in the Yale New Haven Hospital, Elie says his parents were very supportive and encouraged his studies. He also recalls several good high school teachers. He remembers less what they taught than their personalized encouragement. His high school physics teacher gifted him an audiotape of the moon landing, and, upon his graduation, advocated in a letter to his parents that Elie had the potential to be a world-class physicist. Elie was always interested in the natural sciences; for instance, he recalls collecting plants during recess in third grade. He confesses that he was not terribly good at math and found arithmetic rather frustrating and tedious until he got to algebra and geometry, which fascinated him. He was, he says, "*conceptually* good at math."

When it came to selecting a university, he had a difficult choice to make.

Should he go to Yeshiva University (YU) in New York City or MIT in Boston? He decided on the former and did not regret his decision. At YU, very few students majored in physics. In the higher-level physics courses, there were often fewer than six students in a class. For some advanced electives, such as number theory, there were only *two* students in the class, which made such courses feel more like conversations rather than lectures. He developed personal relationships with his professors and had no need to compete for research opportunities. He spent one summer doing computational physics with a professor from Boston University, and another exploring differential equations related to liquid surfaces. He enjoyed working with a team and being invited to accompany his professors to a conference.

For his graduate studies, he went to the University of Connecticut in Storrs. There he found the university and the classes to be much larger than his previous experience. He could see that his theoretical foundations were good, but that he needed more work in topics such as electrostatics and statistical mechanics, so he enrolled in both undergraduate and graduate courses simultaneously to catch up to his peers. Two events in graduate school had a formative impact on his career. He missed the final exam in his first course in quantum theory due to the birth of his son. He called his professor from the hospital to see if he could take the test the following day. The professor denied the request, but said that in lieu of the exam, Elie could do an independent study, writing an essay on the topic of his choice. While working on this essay, Elie fell in love with the foundations of quantum theory. It was pure serendipity, he says. He would never have done a deep dive into the subject, and discovered his passion for it, if he had been able to take the exam. It was most fortunate that his son was born that day!

Later, feeling temporarily dejected by what he perceived as the relatively superior command of math and physics of his peers, he made plans to exit academia with only a master's degree, and had the intention of subsequently pursing law. He thus took the entrance exam for law school. However, one of UConn's requirements for the master's degree in physics, notably, requires the candidate to present on a topic of their choice and be quizzed on it by a panel of professors. Elie selected Bell's theorem, and following his presentation, much to his surprise, one of the faculty members strongly asked him not to leave graduate school. She said it was a great presentation and that she would be willing to act as his thesis supervisor on whatever topic he chose. This was the second event that had a great impact on his career. He did remain at Storrs and worked with Dr. Susanne Yelin, who encouraged him to follow his interests.

As he subsequently pursued his PhD, he repeatedly encountered thought-leading research emanating from the Perimeter Institute in Waterloo, Ontario. Sometimes he would read a paper from Perimeter and be unable to sleep that night due to the exciting potential it contained. Perimeter's outsized impact on his work made it like a shining star to him. He was invited to go there for a month, an experience that reinforced his admiration of the institute. He was particularly interested in Robert W. Spekkens's work that shed light on quantum theory from a perspective of causes and effects. Not only did that work resonate with him in isolation, but when Rob gave a talk to the weekly quantum seminar series on YouTube, Elie found it remarkably engaging and sought to follow up with Rob directly to speak about the work's possible future directions. They did have a conversation, and Rob evidently thought the conversation worth continuing, as following his thesis defence, Elie was invited as a postdoc to Perimeter, his dream come true! From Elie's perspective, there is nowhere in the world that can match what Perimeter offers as a place to conduct foundational research. Perimeter collects the most innovative researchers, both as students and faculty, and furthermore promotes interaction with foreign collaborators by frequently hosting visitors and by generously funding travel. It exceeded Elie's expectations, and he continues to work with Rob, no less enthusiastically.

Elie's advice is that there will be considerable space for inquiry and research for many years to come. While admitting that he may be biased, he perceives vastly more opportunity for young researchers to find their niche in less developed fields such as quantum foundations, as opposed to more established fields such as particle physics or cosmology. His long-term goal is to be able to provide a clearer understanding of quantum theory; like other researchers in his field, he sees his work a bit like unscrambling an omelet. When asked about interdisciplinarity, he sees a rich symbiosis between the subfield of statistics known as causal inference and quantum theory. Indeed, he thinks there are deep connections yet to be made, but that to achieve them researchers will have to overcome terminological barriers in order to better communicate across disciplines. He values time to spend reading and thinking.

Elie is most proud of his work on inflation techniques for causal inference, which was motivated by his need to characterize the differences between causal classical and quantum causal models. Unable to find an existing solution to this problem in the literature, Elie and his colleagues spent nearly two years developing a radically unconventional approach to the problem. Although they had only set out to identify *some* features that distinguished quantum from classical models, it turns out that their inflation technique *completely* describes classical models with unobserved factors. When Elie first presented this work at an external conference, it was met with incredulity. How could he possibly have solved the holy grail of causal inference? Elie's biography provides the answer. He is patient, brilliant, hardworking, and relocated to the place he calls a physics researcher's dream: "Camelot" (aka the Perimeter Institute). He also teaches at Waterloo University, and after nearly a decade of work with colleagues including physicist Robert Spekkens, Elie is highly motivated to continue his work in quantum causal inference, which has implications for a vast array of fields, including AI, banking, insurance, climate research, public health, and epidemiology, for example. Together with other brilliant researchers at Perimeter, they have made Canada an internationally recognized center for their field, attracting students and scholars from around the world to join in their extraordinary work, which will deepen our understanding of quantum models, providing untold opportunities for the development of applications that will have the potential to improve our lives. As the future unfolds, I am sure we can all applaud Elie for caring about his family and physics!

31

STEPHANIE SIMMONS

Building a Large-Scale Universal Quantum Computer

Vancouver, BC

Stephanie Simmons was born blind, and if not for the Canadian health-care system, her family could never have afforded the three surgeries in her first year of life that enabled her to see. Now, as an internationally recognized and sought-after expert in quantum technologies, she decided to return to Canada because she believes she owed a lot to the public education system that trained her and the health services in this country that saved her vision.

Stephanie is grateful to her parents for their relentless love and support. She believes human beings are the happiest when they pour themselves into meaningful work. She herself draws meaning from her research, and hard work provides her much satisfaction.

She was reading at eighteen months and did math at an early age as well. She understood she was gifted and felt it her duty to do the most she could with what she was given. At sixteen she was building websites and programming computers, when she discovered quantum computing, which she thought was very cool. She saw a description in the local paper in 2001 about the newly formed Institute for Quantum Computing at the University of Waterloo, very close to her home in Kitchener, Ontario. She little suspected that quantum computing would become her life. Indeed, she was busy being educated by her parents, who introduced her to things that were not in the school curriculum, from coding to politics. Stephanie was good at math and liked both her high school math and art teachers. She thought she might go into art, as "it influences [her] thinking. It provides [her] technical creativity and the ability to identify patterns across disciplines." She was also a swim teacher and lifeguard. However, when it came time to decide seriously about university, the decision was easy. She said the only quantum institute in the world at that time appeared to be in Waterloo. During her first week at university, she says she naively knocked on the door of the director of the institute and asked for a meeting. She sought his advice. She was registered for computer science, but thought perhaps she would need more math. He agreed. She also wondered if perhaps she should do both pure mathematics and mathematical physics. She ended up being the first person to obtain both degrees simultaneously.

Stephanie's next problem was to decide where to do her graduate studies and in which quantum computing subfield. By the time she completed her undergraduate degree it was evident that the large-scale quantum computer that she dreamed of programming wasn't yet available and wasn't going to emerge for some time. She therefore switched from software to hardware: she headed for Oxford to do her DPhil in materials science, to research the (then) best effort to build a working scalable quantum computer. She was offered a tenured faculty job at Oxford upon graduation, but instead stayed on at St. John's College in Oxford as a Glasstone Research fellow. She had other longer-term plans. She had developed a clear picture in her mind of what needed to happen to build a large-scale universal quantum computer. She spent two years as an electrical engineer in Australia, which was, at the time, the global epicentre of silicon quantum technologies. Then she returned to Canada, to the physics department of Simon Fraser University in Vancouver, which she identified as the best place to try her own ideas for a scalable quantum computer. There she was named Canada Research Chair in Silicon Quantum Technologies and Quantum Computing. She knew that, if her quantum solution were to work, she would have to work alongside Mike Thewalt—a global legend in the optical study of defects in silicon—to develop it.

Stephanie's work has been widely published. She has won *Physics World*'s Top 10 Breakthrough of the Year award in 2013 and in 2015, and is one of only five people in the world to receive this award twice. She received the YWCA Women of Distinction Award and the Arthur B. McDonald Fellowship in 2022. She says that she has been far more successful in developing her quantum technology than she thought possible. It is "beyond [her] wildest dreams." There are very few mathematicians who are now leading large-scale quantum computing efforts, and she attributes much of her current success at full-stack quantum technology co-design to her unique technological background. She credits the patient exploratory research she did for many years with the success of her current results.

Stephanie's interest in quantum computing technology centres on what it can produce of value for people, for example the ability to enable simulation work on climate crisis challenges. She is founder and chief quantum officer at Photonic, a company with 120-plus employees commercializing her technology, but she says her research never ends. She asks, "How many times has a branch of physics been commercialized in history?" Fortunately, she answers her own question: "Many times, but typically not more than once in a generation. Electromagnetism, optical physics, nuclear physics, semiconductor physics: these each now fuel engineering departments worldwide and represent massive, incredibly useful industries."

When asked about the aha moment vs. patient experimentation over time, she says her research life has been a combination of "ahas" and "slow build." She says, "At some point you realize that you've already solved a tough problem, learned a difficult concept, or you actually created something. Before that happens you need to soldier on over and over until you finally recognize that the solution was actually moving into place. Sometimes the solution is there before you can articulate it because the subconscious, which does so much, is always working."

Stephanie's current goal is to play out the commercialization of quantum technologies. Her entire career to date has been dedicated to this, and she is excited to actually use the tools she has created for everyone's benefit. She notes that what could not previously be simulated in classical computers will now be possible on quantum computers and that many doors are being unlocked, enabling more research. It is hard to imagine the limits for quantum applications. Some things are easily predictable. "When transistors were developed, it was easy to foresee their application in hearing aids.

Other things are harder to define. When computers were developed, did anyone imagine Facebook?"

Stephanie deems travel to be important to learning and thinks that Canadians have a bit of an inferiority complex. When you live overseas you get to feel a bit what immigrants feel. We all need to develop a broad base of trust in others. We are all people, and we are all the same in the end.

The advice she would give people considering research in quantum is this: "Human beings tend to regret risks not taken. They do not regret taking risks, even if they end up failing. So, when in doubt, take the leap of faith! Everyone suffers from self-doubt and everyone, men and women alike, suffers from the imposter syndrome far more than you'd think. This is self-limiting, so just take it as a given and get on with work—just do it!"

Stephanie has at last found the technology she sought for the last twenty years. Now she hopes to make a difference in the world. The consequences are "too wild to imagine." She thinks she is so fortunate to be back in Canada and at Simon Fraser University and to have taken the time to talk. It has given her the opportunity to reflect on the number of things that had to "go right and measure up." She says she is fortunate to have fine people who followed her, and she is happy to have found a place that supports "really risky stuff." She has been named co-chair of the advisory council on Canada's National Quantum Strategy. We are all doubly fortunate that she returned to Canada to share the results of years of hard work and effort and that she is dedicated to improving the environment and health in the interests of all—through quantum, naturally!

32

NICOLE YUNGER HALPERN

Quantum Steampunk—The Engine That Does Not Exist—Yet!

College Park, MD; Waterloo, ON

An alumna from the Perimeter Institute Scholars International Class of 2013, Nicole Yunger Halpern's quest is to reinvent thermodynamics (the science of energy) for the quantum age, and the title of her book is *Quantum Steampunk: The Physics of Yesterday's Tomorrow*.

Her research helps develop a modern theory of thermodynamics and harness the theory to transform other spheres of science. The book in which Nicole asks profound questions frames them within a bit of a sci-fi adventure set in Victorian times, the days when steam engines were invented. She then describes a quantum engine, which of course does not yet exist, but the seed has been planted in our minds.

There is no need to inquire if Nicole believes in interdisciplinarity. She lives it and works at the crossroads of quantum physics, information science, and thermodynamics. She regularly consults colleagues in high-energy physics, optical physics, condensed matter, and chemistry. Her book draws a parallel between the science behind steam engines powered by the energy extracted from heated gases and the work researchers now do to understand quantum entanglement, for example. Nicole notes that quantum thermodynamics shares its aesthetic with art, literature, and film (the steampunk genre), and she says that every field of science has its own aesthetic. More and more, there is an entire culture of people who connect science to art, literature, and film.

Born in Queens, New York, Nicole and her family moved to Florida, which was "a lovely place to grow up," Nicole affirms. Her parents' priorities for her were that she would be happy and make a living. She loved studying everything and, in particular, she loved to read, even over her brother's shoulder, and would bring a book to read while waiting for her order in a restaurant. She had a broad education and benefitted from English and Spanish literature, philosophy, history, physics, and computer science. She enjoyed building worlds in her mind and can "draw useful connections that people might not usually make." At one point she thought about becoming a librarian. She had done an internship at the Smithsonian Libraries, and while there studied the list of manuscripts and books from the Dibner collection of rare books and manuscripts, and she was later able to read many of them. On her first day on the job, she was handed a first-edition copy of Galileo's *Sidereus Nuncius* (*Starry Messenger*). This was indeed a privilege. Nicole still has great admiration and respect for the work of librarians, but has opted to be one of the people who comes to them for help with a project.

When she arrived in college she looked for more answers to questions and is still "studying some of everything." However, thanks to her wonderful physics professors, she saw physics through other lenses and could see linkages to mathematics, history, and philosophy. She put together a special major combining these fields and convinced the department chair that she had already completed the equivalent of these courses. She took her first course in quantum computation and loved the mathematics behind the theory, which she also studied in an individual-directed reading course. She was certain she would like to go to graduate school and thought about dedicating her life to this subject and spoke with some faculty members to obtain advice.

She wondered if thermodynamics would help her understand quantum information theory and went to the Perimeter Institute in Waterloo, Ontario, for her master's, where she wrote her culminating research essay on quantum information. She then went to Caltech in Pasadena, California, for her PhD. Not finding anyone at Caltech who was doing quantum thermodynamics, she spent some time in Oxford, where there was a group working on the topic. Her goal was to do the most creative work she could at the intersection of quantum information theory and thermodynamics.

Nicole wondered at that time about the difference between quantum and classical physics. Quantum physics is known to enhance information processing, and that is why scientists and engineers are building quantum computers, networks, and cryptographic systems. Just as information can be processed, so can energy. So one might expect quantum physics to transform our understanding of energy processing, as happens in quantum thermodynamics. One landmark Nicole accomplished in her doctoral dissertation was to prove a thermodynamic equation that describes how information is scrambled when it falls into a black hole. The equation prompted a new way of measuring information scrambling in experimental systems intended to model some features of black holes. She wrote a paper about loops backward in time, looking at chaos theory, thermodynamics, and black holes.

"Chaos is," she says, "really about our inability to reverse time." She also wrote a book on quantum thermodynamics for the general public. She brings together the old and the new and opens the door to possibilities for the future.

Nicole, who was a member of the outdoors hiking and camping club in college, says walks are good breaks from poring over papers and allow for processing the science. Nicole aspires to pay forward the mentorship from which she has benefitted and is delighted to hear her students and postdocs brainstorming about future directions for their research together.

Nicole loves research and says that she has about one hundred colleagues researching quantum topics at the National Institute of Standards and Technology and the University of Maryland, where she is now a professor. At the conclusion of her book, she writes: "We've translated theoretical proposals into experiments and partnered with other fields of science. I expect quantum thermodynamics to continue thriving and evolving. Where the past meets the future, as when thermodynamics meets quantum computation, science can spin a today fit for a steampunk novel."*

Nicole's advice to people who want to study physics is "Do not fret if you are not having fun every minute. Tough integrals frustrate all of us.

---

* *Quantum Steampunk* (Baltimore: Johns Hopkins University Press, 2022), 255.

But physicists have the opportunity to learn and grow every day, so enjoy the ride." Nicole admitted on the Perimeter Institute website that she is "terminally interested in everything." This curiosity will certainly lead her to discover new connections that will enhance our understanding of the flows of time and energy. Nicole is following the leads her inquiring mind establishes, be they in the science of the past, or of the present. The result will doubtless be new avenues for science, new fields of science. Her story is one of innovation writ large, and intelligibly by a woman who studied and was inspired in Canada.

PART FIVE

Machine Learning and the Sense of Metaphor

Today, computers react faster to swings in the stock market than experienced traders on the floor. They can handle large quantities of data and analyze it, and rapidly pick up trends and inconsistencies. The next logical step in the evolution of the computer has been to draw conclusions and incorporate the information (and perhaps misinformation) gleaned from the web: "self-teaching" in a way. This is what is often called machine learning or deep learning and is the current challenge that engages researchers around the world.

Artificial intelligence, a relatively new development in computer science offering exciting possibilities, encompasses overall computing systems that can perform tasks like problem-solving, learning, and planning. Machine learning is one aspect of AI and refers to the ability of computers to process large quantities of data rapidly and to draw conclusions from the data based on the parameters that have been programmed into the system. The programming can include instructions for "learning" so the conclusions will improve with the quality, amount, and characteristics of the data submitted. If the data is language, the program must be extremely complex and agile at the same time. For example, a word may have more than one meaning or can be employed sarcastically or ironically. The machine needs to be programmed to recognize all usages and all contexts. Many of us recall having been told by our parents that certain words were not to be "used in this house!" Other words are reserved for the workplace, others for poetry, and yet others for that special occasion when we might meet royalty. The machine needs to be programmed to respect etiquette and context.

Then there is the sense of metaphor. One of my earliest memories occurred when I was four years old. I was trying to push open a door and it was not cooperating. My mother said, "Use your head." So I did exactly that. Luckily, I avoided a concussion, but gained two things: a small bump on my forehead and an understanding of metaphor. It was as if a whole universe had suddenly opened up to me. Words could have more than one meaning! Language contained secrets to discover!

Now, had I been a computer I would have continued to bash my head (neural networks) until I could be reprogrammed to recognize the double meaning of that word. It becomes quite a job to think of every linguistic variant, especially since language is constantly evolving. What was "hot" one day is "cool" the next. Worse, if a computer is programmed to acquire new words, it will, if exposed to social media, unfortunately "learn" many expressions that should never be repeated, that might be prejudicial, misogynistic, or racist.

If artificial intelligence is applied to problem-solving, one might program a computer tasked with driving a car to stop when an object appears in front of it. Embedding this feature has been demonstrated to be very effective. The machine can react promptly and often faster than a human driver. On the other hand, when the computer causes the car to stop too promptly, it might cause a rear-end collision with the car following behind. Unlike the human driver, the computer will not slow down because it is tired. However, it might not make that split-second decision to swerve off the road instead of braking to avoid a worse accident. The human creator of the program must try to think in advance of every eventuality and program the computer to make life-and-death decisions. Sometimes these decisions involve ethics. If the object in front of the car is a skunk, is it better perhaps to simply continue moving forward rather than stopping short or driving into the ditch? We might use math to rapidly calculate the speed of the object ahead and the angle at which it might be encountered to result in the least damage. These complexities must be considered.

Planning is another area where computers already look at masses of data and tell us that if we want to travel from point A to point B right now, it will take twenty-nine minutes. If it is Monday at 7:30 a.m., it may take a bit longer; and if it is Sunday at 7:30 a.m., a bit less. The more complex the system, the more variables that may have an impact on driving time. A computer might be programmed to take into account holidays, weather, and construction, for example.

The ability of machines to learn, the observations they will make, and the conclusions they will draw are the responsibility of humans who design the programs that provide the basis and the context, the rules and parameters for analysis and conclusions. Machines do not have consciences. They do not actually think. We program them and the current challenge is to program them for the good of humanity by ensuring our programs include ethical considerations.

Artificial intelligence and machine learning have three parts: the idea or concept, the machine, and the program. These must all be governed by overarching ethical values that protect and promote humanity: enabling individuals and communities to enjoy peace, good health, and prosperity.

Known as two of the three godfathers of artificial intelligence and machine learning, Geoffrey Hinton and Yoshua Bengio have led the development of deep learning, which enables computers to recognize speech and handwriting and to generate natural language and translations. They create algorithms that will allow artificial neural networks to solve problems. The applications are endless and include diverse fields such as robotics, banking, and health care. Raquel Urtasun's work in creating the simulations that will enable driverless vehicles to solve issues of supply and demand illustrate one such application. They agreed to share some of their inspired thinking and their care and concern for the ethical deployment of the possibilities they are creating.

33

GEOFFREY HINTON

On Algorithms for Deep Learning and Those Aha Moments of Innovation

Toronto, ON

Geoffrey Hinton pauses to reflect on moments that led to his stunning breakthroughs. Finally, he tells me that he got one of his best ideas waiting in line at the local branch of his bank. He was a bit annoyed with the bank because each time he got to know a teller, they would disappear and a new one would appear. Geoffrey wondered why the tellers were always being moved around. He was sure there must be an organizational principle.

After giving it some thought, Geoffrey guessed that the reason for the change in tellers was to prevent them from conspiring to steal from the bank. He then drew an analogy from the bank tellers to the noisy behavior of neurons, which are nodes that transmit information in a neural network within the computer. They often act as if they were absent (like the bank tellers). This suggests that the reason that neurons drop out (act as if they are absent) is to prevent "conspiracies" between neurons. If a neural network can rely on all its neurons being present, it can rely on some neurons to correct mistakes made by other neurons. At first blush, this seems effective. But such a network is actually very fragile, as all neurons are needed, even to deal with slight changes in a task. However, if neurons disappear at random, the solution learned by the neural network is much more robust to small changes. This is what happens in a bank when one person is missing and the others cover for them.

This same phenomenon was also discovered by evolutionary theorists. If animals simply clone themselves, they cannot cope with big changes in their environment, but if you always replace half of the genes from another individual, they cannot rely on co-adaptation of large numbers of genes, so they become much more robust to environmental changes. Thus, the same phenomenon explains why neurons are noisy and why reproduction is sexual.

Geoffrey not only uses analogies to explain his ideas, but he is also inspired by analogies to consider how we can organize networks to repeat tasks and to deal with unusual, singular actions as well as larger numbers of tasks.

Geoffrey crosses disciplinary boundaries, combining art and science, computing and psychology. "Look at things that are apparently different, but then find what they have in common," he says.

The recipient of the A. M. Turing Award, also known as the Nobel Prize of Computing, Geoffrey Hinton is, with computer scientists Yoshua Bengio from Canada and Yann LeCun from France, one of the three godfathers of artificial intelligence, or deep learning. He works on artificial neural networks, which are designed by combining computer science and statistics to simulate the behaviour of the human brain in problem-solving.

Geoffrey is most unusual in that he is able to understand the mechanics of the computer and the functioning of the human brain. His work crosses the disciplines of psychology and computer science. Born in 1947 in London, England, he moved with his family to Bristol at the age of one and a half. There, he grew up in a family that included a great-great-grandfather whose work in mathematics has been considered to be one of the foundations of computer science. As a child, Geoffrey loved math and algebra, but in later years he did not have a great affection for advanced mathematics and had a particular distaste for functions. He jokes that this was probably because he was not good at math beyond algebra (his students can't believe this). He offers as proof the fact that he decided to study psychology in his final year of undergraduate studies at King's College, Cambridge, and he recalls his professor in mathematical psychology, David Ingleby, as being particularly helpful in thinking across disciplines. On completing the program in 1971, he took a break and worked in carpentry. After a year, he decided he was not interested in pursuing carpentry as a career, though to this day he still maintains an interest in it as a hobby. He decided that he would rather study artificial intelligence. He had learned about a new program in artificial intelligence at the University of Edinburgh from his Cambridge classmates and thought it might just be easier than carpentry.

In 1975 Geoffrey completed his PhD at Edinburgh and, on graduation, discovered that there were no jobs in artificial intelligence in the UK. He worked as a research fellow in the cognitive studies program at Sussex University from 1976 to 1978, when he received a fellowship as a visiting scholar at the University of California in San Diego, where he could continue his work on neural networks in the company of excellent researchers and in a warm climate. He returned to England in 1980 as a scientific officer in the applied psychology research unit in Cambridge. But after two years, he went back to the US as a visiting assistant professor of psychology for six months at UC San Diego. He was then offered a faculty position at Carnegie Mellon in Pittsburgh, Pennsylvania, where the resources and the graduate students were both very good. However, after five years, Geoffrey decided to leave for political reasons. It was during the Reagan era, and he preferred not to receive military funding for his research. He was then hired as a professor of computer science at the University of Toronto in 1987, but in 1998 was attracted back to England to be founding director of the Gatsby Computational Neuroscience Unit at University College London. In 2001, he was offered support in Canada from the Canadian Institute for Advanced Research (CIFAR), which, combined with a position as Canada Research Chair at the University of Toronto, gave him sufficient time to dedicate to his research and graduate students. He moved to Toronto, where the lab equipment and computers were not as good as those in the US, but he received a grant from the Canada Foundation for Innovation, enabling him to be one of the first machine learning researchers to acquire Nvidia's graphics processing units, which were the best computers in the world for his work and greatly advanced his projects. In addition, he decided that Canada was a better society in which to raise his children.

Until recently he divided his time between Google, where he worked at Google Brain, the AI research team, and the Vector Institute for Artificial Intelligence, a not-for-profit corporation that he co-founded with computer scientist Raquel Urtasun. It brings together academics and industry-leading research in machine and deep learning.

Geoffrey believes that it is important to have people with whom you can discuss ideas and new concepts. After all, his inventions are not all easily understood in casual conversations with neighbours, for example. He has therefore always stayed in touch with colleagues from the places where he has worked. At Carnegie Mellon, he established, with a colleague and funding from the Sloan Foundation, what was called the Connectionist summer school to bring together graduate students from around the world who were working on neural networks to meet and share ideas. He carried on this concept over the years, organizing similar workshops for CIFAR. He even taught a free online course on neural networks for Coursera.

Geoffrey's innovations are many and include co-inventing the Boltzmann machine (recurrent neural networks where nodes make binary decisions, useful for machine learning) with David Ackley and Terry Sejnowski, and contributing to Helmholtz machines (a type of artificial neural network that can be trained and account for a set of data), the products of experts (a machine learning technique), distributed representations (when a pattern of activity is represented by one network and is represented by another across processing units), time-delay neural networks (the wake-sleep algorithm), and capsule neural networks, which he says are "something that finally went well." He says he has had many aha moments, but also many downs, and that nearly all his innovations come from seeing that things that are apparently different have something in common.

Geoffrey's biggest breakthrough was perhaps when he realized that a quantity used in statistical physics ("equilibrium-free energy") is mathematically identical to posterior distribution, which is a way to balance newly observed data with prior knowledge and is used in statistics. Understanding that they are basically the same led to new training models and exciting innovations, both small and large.

Geoffrey never stops thinking, and in 2022 he introduced a new learning procedure for neural networks.

When asked about the challenges he sees before us, Geoffrey says we need to develop new kinds of computers that function at lower power, and he looks to the complex analog computers that were used for complex calculations in the 1970s and have since gone out of fashion. Analog computers use continuous values and thus resemble more closely the function of the human brain, while digital computers use discrete values. We need to learn how to train the purely analog computer system, but nobody knows how to do this—yet!

His advice to those considering a career in research is to never give up—even if others tell you that you are wrong. Pursue your ideas until *you* know that they are not a good idea and, when that is the case, move on quickly to the next idea. He also recommends embracing the possibilities offered by AI that could relieve us of many mundane, arduous, and even dangerous tasks, making our lives better and more enjoyable.

Geoffrey did not, however, think the recent advances in machine learning would occur as rapidly as they have. The power and capacity of machine learning have given him pause. He stepped back from Google to ponder the possibilities, limitations, and dangers. Before we fully open Pandora's box, he suggests we stop and take time to reflect on ways to protect the interests of humankind. As computer scientists around the world contemplate responsible and ethical guidelines, regulations, and controls for further programming and implementation of machine learning, Geoffrey considers the potential and the danger that lie before us. Our future is in good hands and a great mind.

34

YOSHUA BENGIO

Deep Learning, Generative Models, Applications, and Ethics

Montréal, QC

"The brain is an idea machine, and it works all the time," says Yoshua Bengio, who is considered one of the three godfathers of artificial intelligence and is the winner of many awards, including the A. M. Turing Award for Computing (with Geoffrey Hinton and Yann LeCun for their research in deep neural networks), the Killam Prize, the Herzberg Medal, the Order of Canada, and France's Légion d'Honneur.

Yoshua was born in Paris in 1964. He and his family moved several times during his childhood, ending up in Montréal, Québec, in 1977. He enjoyed school, especially mathematics, and the sciences, as well as languages, and never wanted to miss a day, even for a family holiday. He and his brother shared a paper route, and they used the money they saved to purchase an Atari, one of the early computers that worked with tapes, but allowed them to explore programming. Later, they bought a programmable calculator, and around the age of fourteen or fifteen, Yoshua taught himself to create programs using simple, basic assembly language. His passion for computing grew quickly and, wanting to learn more, he took a course in his CEGEP, which was disappointingly boring. He did, however, have a math teacher who inspired his continued interest in the subject. "To this day, mathematics has had a strong influence on my work," he says. "Math is an amazing tool."

When he went to McGill University in 1986, Yoshua thought of studying physics, but opted instead for computer engineering, and then went into computer science for his graduate degrees, which were both in computing science, at McGill University (1988, 1991). Yoshua believes that we can train our minds to think innovatively. Neuroscience has demonstrated the plasticity of the brain, so innovative thinking is something we have to practice. We need to ask ourselves how things work, how to fix things, and how to find solutions. When something does not work, the scientific mind will ask why and the innovative person will seek solutions. Yoshua has led research in speech and handwriting recognition, natural language generation, and machine translation. He is fascinated by generative flow networks that employ energy-based probabilistic models to reinforce learning. His GFlowNet is, for example, able to sample a diversity of solutions and follow a series of constructive steps to identify probability. Yoshua works to understand the mechanisms that give rise to machine intelligence and to imagine applications to other fields like cell biology. "The mind is amazing in its ability to find analogies," he says.

During the COVID-19 pandemic, Yoshua asked himself, "What can I do?" He became interested in epidemiology, specifically in how cells work and break down. He wanted to see if it would be possible to speed up drug discovery through the design of experiments, the automatization of processing, and the faster interpretation of data. He worked with colleagues to support pharmaceutical production and is still scientific advisor to a major firm. He is now thinking about anti-microbial resistance, an area that requires investment by industry and government to provide incentives for researchers to step up their work.

Yoshua separates mission-oriented research from curiosity-driven research. The first has a goal, a problem to solve, and the central question is "How?" Metaphorically speaking, mission-oriented researchers are looking for more efficient ways to hammer. On the other hand, the goal of curiosity-driven research is discovery and the central question is "Why?" These researchers have a hammer and are looking for the nail.

Yoshua has had quite a few aha moments. These were not generally major discoveries, but breakthroughs in understanding that were extremely powerful emotionally, a great personal pleasure. Indeed, he muses, as we grow older these moments are perhaps the only pleasure that remains strong. However, he cautions that these are not "magic moments," but the result of long and careful thinking. He never stops thinking, even during his daily walks. Sometimes he is awakened in the middle of the night by an idea that could potentially solve a problem on which he has been working.

Yoshua believes in global partnerships for international solutions. He notes that Canada and France have worked together in AI and that other countries are now joining the collaboration. There are also annual meetings where researchers in AI come together from around the world to share ideas and find the questions they should collectively consider.

Yoshua notes the large gap between the current abilities of computer intelligence and human capabilities. Current AI is good at perception but not at reasoning, whereas classical AI was good at reasoning but not perception. When this gap is closed, it will have a big impact on AI systems and their ability to adapt and avoid catastrophic errors. The system will have to understand its own limitations. Conscious processing does not currently exist. Yoshua thinks that might take another fifteen years. When it does, it will transform AI.

Yoshua also points out that there are innovations that have clear and fairly immediate commercial value and others that have great value in fields like health care, for example, where the impact will save lives by making the system more efficient and better able to serve the population. He says that while we admirably tend to support the innovations that have an impact on the financial market, we "leave a lot of innovation on the table." He hopes that we can consider the value of these innovations and the risks to the population if we do not encourage, invest in, and adopt them.

His advice to future innovators and researchers is to "find your own way. Be self-motivated. Keep thinking—when you are walking, when you are in the shower, when you are dreaming. When your eyes are closed, listen to your thoughts. We are fortunate to have a culture of freedom and liberal democracy, where each person can find their own way and explore personal interests. You must be patient and you must first study. You cannot be superficial. You must become an expert, and to do this, you must keep asking yourself, 'Why?' You must not be satisfied just to read words on a page. You need to understand what people really mean."

Yoshua says our youth are not trained to think analytically and critically in schools. They are taught how things were and how they are. They need to be taught to ask questions, to construct scenarios with possible answers on their own.

Yoshua's ideas in machine learning have revolutionized computer science and our lives, enabling the application of AI to banking, transportation, medicine, robotics, and other fields. By sharing his ideas with his students at the Université de Montréal, where he is a professor, he is ensuring the future of his field. By working with industry, he is speeding up the application of his ideas. He has co-founded a company, Element AI; joined a start-up, Botler AI; and he is the founder and scientific director of Mila, at the Université de Montréal's new campus, where a thousand researchers from universities and industry specializing in machine learning come together to promote science and innovation in AI for the benefit of all.

There is an additional aspect to his innovation that will never be forgotten, and that is celebrated the world over: the Montréal Declaration for Responsible Development of AI, a statement defining and supporting the ethical development of deep learning that researchers, universities, agencies, and companies around the world are endorsing. Indeed, Yoshua's leadership in this endeavour has resulted in Canada's international recognition as a nation with high ethical values.

Yoshua says that, years ago, computers were simply seen as "cool gadgets." People did not necessarily think about their impact on society. Now we see them as powerful tools that can be used for good or bad purposes. They might also have unintended effects on humanity. Thus, he reasons, it is essential for all innovators to be thinking about and to have information and knowledge about how society works in order to figure out what might go wrong. Scientists must train their students to be aware of all the possible outcomes of their work. He takes heart in his belief that the younger generation cares more about such consequences than those who came before them.

Travelling across the country, I myself have heard researchers refer with great pride to the Montréal Declaration. When I have asked them about their hopes for the future of research in Canada, many have said that they would like Canada to be known as a country where research work and researchers are ethical and where the population cares and values thoughtful leadership. A commitment to ethical values is essential if we are to be able to employ new technologies, including AI, in solving problems of health, our environment, and the economy.

Yoshua is a brilliant researcher in machine learning. He has invented new ways to improve the technology and he has also thought about the potential impacts of AI. Yoshua offers an inspiring example of true leadership and innovation!

35

RAQUEL URTASUN

Going for Infinity Percent and Driverless Vehicles

Toronto, ON

A world-leading expert in artificial intelligence for self-driving cars, Raquel Urtasun recognizes Toronto and the University of Toronto as the "best place to be for AI." She cites the university as an "amazing, vibrant place" and applauds the exciting environment in a city that welcomes diversity. She also notes that having colleagues she admires like Geoffrey Hinton is key.

A native of Pamplona, Spain, Raquel loved math and physics, but also enjoyed literature and sports as a young scholar. As an undergraduate at the Universidad Pública de Navarra, where she earned her bachelor's degree in 2000, her first research project was in AI. She was impressed by its potential to transform our thinking and our ability to achieve scientific advances that would change the world. She also found it mathematically fascinating. It was, she is quick to admit, "love at first sight."

In 2006, she completed her PhD at the École polytechnique fédérale de Lausanne and spent three summers at the University of Toronto while working on her doctorate. She then did postdocs in Boston at MIT and the University of California, Berkeley, taught at the Toyota Technological Institute affiliated with the University of Chicago, and was also visiting professor at ETH Zurich in Switzerland during the spring semester of 2010.

As an undergraduate she worked with a faculty member in his lab and was soon convinced that she would one day be a professor as well. She is delighted that she did just that because she loves seeing the potential in students and encouraging and supporting them. She joined the University of Toronto in 2013 and became a Canada Research Chair in machine learning and computer vision from 2015 to 2017. She then joined Uber ATG as chief scientist and head of R&D, a position she held from 2017 to 2021, returning to her position as full professor in the Department of Computer Science at the University of Toronto. She acknowledges having shifted her attention to industry, but only after asking herself, "What are the real problems? On which of them should we focus? What are the ones we could solve with the resources we have?"

When she was chief scientist at Uber, which offered her a new perspective, she had feared that she would lose the freedoms she enjoyed as an academic, but found instead that she had more freedom at Uber, where she could be more entrepreneurial and the technology she had worked on for the last decade would get to market faster.

In 2020, she founded her own AI company, Waabi, which is building the next generation of self-driving technology, in response to what she perceived as a broken supply chain. Goods were not arriving quickly to consumers. It was, she said, "a problem of logistics, which could be alleviated with self-driving trucks, with the appropriate safety protocols." Trucking and taxi services would not be limited by the number of drivers available. Vehicles could be more environmentally friendly (electric, for example), better, faster, less expensive. There would be fewer traffic jams and definite benefits for all.

Raquel says the future is bright, and there are many opportunities for young researchers to shine: "We all need to reach for higher goals and push ourselves to improve. We can become our better selves." We need to work continuously on self-improvement, and we should "not be shy to try. . . . Have no fear of failing and you will do well."

Raquel thinks that diversity is important, and people with different backgrounds—be they cultural, racial, or disciplinary—challenge our assumptions.

When she is working, she often relies on intuition, which means that she will pursue a problem by following a certain path she chooses. She sees innovation like a series of steps in your mind. When she was a young researcher, she went to almost every possible talk on AI. Afterward, she would ask herself, "Given what you have heard, how would you do it better?"

Raquel aims for significant change. She thinks that sometimes students are encouraged to publish a number of small papers early in their careers. She thinks that it is perhaps more important to wait until one has a significant contribution to make. She also thinks that innovation will be realized when companies and academics are brought together. She believes companies in Canada do not realize what academic research can do for them. A woman of action and a woman of her word, Raquel co-founded the Vector Institute with Geoffrey Hinton to build a bridge between business and academic research. She thinks the results Vector has achieved in Toronto are proof not only that can the bridge be built, but that it will attract talent and enable the meeting of minds that, in turn, will engender true innovation. She sees the model being replicated across Canada.

Raquel believes that there are two ways to use one's brain. One is the way engineers are taught, which is to move forward by layering steps, one upon the other incrementally. The other is what a person trained to be innovative will do, which is exactly the opposite. They will take giant steps, skipping rungs on the ladder. The two ways of thinking come from different parts of the brain. Both are important and together they will transform the world.

Raquel would like to be sure that everyone who has the ability to be innovative, regardless of where they are born, their race, gender, or ethnicity, can develop their talents and increase their knowledge. She is proud that she has been able to do something different, and her hope is that Canada will become a world leader in innovation.

Raquel has held the Canada Research Chair in Machine Learning and Computer Vision, and has received dozens of prestigious awards, including the Steacie Prize from the Natural Sciences and Engineering Research Council (NSERC), three Google Faculty Research Awards, and a Connaught New Researcher Award, to name a few. She has also been named Chatelaine Woman of the Year.

Raquel's most notable innovation has been Waabi World, a closed-loop simulator for self-driving cars. Along the way she built two bridges: the first between fundamental and applied research, and the second between research and business. She has created an environment that fosters innovation and a climate of high energy, spirit, drive, and passion that will bring the worlds of business and academe together and result, among other innovations, in new improvements to self-driving vehicles. Raquel is highly motivated to push forward the creation of the next generation of transportation. She wants to be a step ahead, but also realistic. To succeed, she says, "You have to go all in, not eighty percent. That means go for *infinity* percent."

PART SIX

The Art of Innovation and Innovating Art

All art is essentially innovative, and the world has long applauded the artists who figured out how to make beautiful glazes on pottery, to create the perfect blue for stained glass, to use new tools to apply paint, or to create music with new instruments. Each artist possesses a creative genius.

Artists have long crossed genres and mixed media. Poetry has been set to music and paintings have inspired poems. Artists have looked to the future while integrating the past. They have done performance pieces where the act of creation is more important than the creation itself. They have involved their audience as spectators and participants. When the performer stops singing, holds the microphone out to the crowd and the audience picks up the tune, the audience becomes a performer and a creator and the sound is truly unique.

Relatively few artists can claim the accomplishments of the talented individuals in this chapter. Tania Willard, for example, takes art out of the gallery and makes nature an integral part of the work. She invites individuals to join art shows not only as spectators but as part of the final oeuvre. Jeremy Dutcher uses music and dance to bring to life a language that is on the edge of extinction. S. Gordon Harwood takes emotional inspiration from the audience and liberates the spectators' thoughts and spirits. M. G. Vassanji writes about a home that no longer exists and creates a new home for others, inspiring underrepresented voices to write and then publishing them. Tina Fedeski and Margaret Maria Tobolowska play and compose while making music accessible to hundreds of young people, crossing linguistic, racial, and cultural boundaries. They create a strong sense of community and encourage young musicians to learn and teach others in turn.

These artists are not introverted souls. They are collaborative geniuses whose innovation lies in the way they integrate us all in their work. Their stories are fascinating and await your discovery.

36

JEREMY DUTCHER

The Magic of Music and the Mysteries of the Past

Montréal, QC

Jeremy Dutcher is a performer, composer, activist, musicologist, and winner of the 2018 Polaris Music Prize, the 2019 Juno Award for Indigenous Music Album of the Year, and the 2022 Library and Archives Canada Scholar Award. His work is about rediscovering and reimagining our relationship with those who have gone before us and with those who will come after us—all in the perspective of the present moment. Jeremy is a bit like a magician, weaving the wisdom of the past together with the future we envision.

A Wolastoqiyik (Maliseet) member of the Tobique First Nation in New Brunswick, Jeremy was born in 1990 in Fredericton. His parents gave Jeremy and his brothers the freedom to explore their talents and ideas and encouraged them to do well in whatever they chose and to be happy. He learned keyboard skills from an older brother, an inspired pianist, sparking his love for music.

Jeremy's interest and sense of his cultural past were derived from early experience. Jeremy's mother was what is known as a "silent speaker." She had not been allowed to speak Wolastoqiyik in school, and while she used it with her sisters and Jeremy heard the beauty of their words, he was more "fluid than fluent" in the language. Growing up, he knew many of the words that his mother would say, including "I love you" and "Behave yourselves," but Jeremy was curious to learn more and eventually made learning his mother's language an everyday priority. He describes it as a journey, stretching and reaching for words.

From a young age, Jeremy also listened to stories of the past told by the elders, who became a source of cultural knowledge, which he later augmented with transcriptions of Wolastoqiyik songs that had been recorded on wax cylinders in 1907. Jeremy visited the archives of the Canadian Museum of History and, transcribing and translating the texts, magically heard the sounds of his ancestral voices. Immersed in the music of the written words, which evoked the rhythms of traditional songs, Jeremy combined them with his own, new compositions that were inspired by the ancestors. This was a radical departure from his studies of classical music and opera. Jeremy notes that at first "people were not sure what I was doing, but I did not know, either. Today I write songs, and orchestras play my music, but I actually wrote the music with my ancestors and simply tried to create a bridge between the classical and the traditional worlds." For inspiration, he cites Tanya Talaga, acclaimed author of *Seven Fallen Feathers*, and remains thankful for her words of wisdom. He is also inspired by the electronic music group the Halluci Nation (formerly known as A Tribe Called Red), whose music sounds very different from Jeremy's, but is similar in that it is a compression of old and new sounds.

In 2013, Jeremy graduated from Dalhousie University in Halifax, Nova Scotia, where he studied music and trained as a classical, operatic tenor. He found the program very "constraining." He notes however that diamonds are made from great pressure, and perhaps the pressure of the music program helped him identify what it was that he truly wanted to do, which was to rediscover his past and integrate traditional songs with classical form and personal inspiration. He remembers hearing Buffy Sainte-Marie say, "If what you want is not on the menu, go to the kitchen and cook it up and show them how good it tastes. So go on and enjoy the cooking of it." He was thrilled to sing a tribute to Buffy Sainte-Marie at the National Arts Centre in Ottawa in 2022.

Upon graduation from Dalhousie University, Jeremy took a job in Toronto at Egale Canada, which is Canada's leading 2SLGBTQI human rights organization. Its work included outreach with Indigenous communities, discussing how to encourage reserves to be more inclusive to people of all genders and abilities. In evenings, he would work on his music. He knew he had to be in Toronto, where there was a "music ecosystem," and he started getting gigs.

At this point, however, he faced a difficult decision, one he calls "the leap." He liked the job at Egale, but he was passionate about his music and wanted to pursue it full-time. He was not sure he could succeed as an artist, but he felt in his heart he would have to choose his art: "I just had to do it. I determined that I had to go and make music." He quit his job, played music—and people listened. It simply snowballed from there. He hired a manager, who built the infrastructure needed to record his music and get it to market, where it was heard across the land. His award-winning album came out in 2018, when it received the Polaris Music Prize, and additionally, in 2019, the Juno Award for the Indigenous Music Album of the Year. The title of the album is *Wolastoqiyik Lintuwakonawa*, and it was recorded in Jeremy's native Wolastoqiyik language. With only one hundred speakers of the language, Jeremy's popular recording made history as the only album to date produced in Wolastoqiyik. It is both a tribute to the past with quietly evocative leitmotifs that sadly recall ancestral songs of days that are disappearing, and powerful, vibrant passages that are a clarion call to the future. Jeremy combines the high notes and drama of the opera along with costumes and incredible energy and stage presence in his performances. Jeremy said, "Canada, you are in the midst of an Indigenous renaissance. Are you ready to hear the truths that need to be told? Are you ready to see the things that need to be seen?"*

Jeremy is currently working on his second album, which may take some time because he admits he is a perfectionist, and the work has to be of the highest quality before he releases it. In the meantime, he is also working with a choreographer on a ballet. They have carte blanche to create a ballet tentatively titled "Medicine Fusion," which will be the first ballet completely in Wolastoqiyik and performed by the Atlantic Ballet of Canada.

Jeremy attributes his innovative creations to thoughtful study,

* See: Sean Brocklehurst, "Deep Listening: How Jeremy Dutcher Crafted His Fascinating Polaris Prize-Winning Album," CBC, September 18, 2018.

understanding and respecting the past, listening and learning, and finding a way to weave this all together. He has worn glasses since the age of eighteen months and remarkably he does not wear them while playing music. Without them, the world is painted in broad, bold strokes, shapes, and colours, and he can sense and feel more. This is somewhat reminiscent of Edgar Degas, the French impressionist painter, who also had poor eyesight that gave him a very special artistic vision.

Jeremy is a time traveler, combining past, present, and future. He works across disciplines, combining history, music, colour, and costume with opera. He listens in his heart to the voices of the past, which he hears as laughing voices, as poetry. He imagines himself sitting next to the elders, and he recalls an elder, Maggie Paul, who told him to "bring songs and dances back to the people and all will be once more as it was once before." He thinks song is powerful and can start a movement that can change the world. Music helps us tell the good stories that have been forgotten by time and buried under the weight of sadness.

Jeremy's advice to young people is "I am allergic to advice. All of us have different paths on which to walk. How could I say anything? But move in love and gratitude and you cannot go wrong, whatever you do. You should always remember for whom you are working. Your community? Yourself? That is good. But *never* for money." And then he smiles and says, "It sounds like I may actually have offered some advice!"

Jeremy concludes with hope for Canada: "Once we [Indigenous people and settlers] were not on the same page. We were not even in the same book. We have come a great distance and we must all live together. We can actually have a conversation today and chart a new course together: a course of kinship, family, and love." With Jeremy's inspired vision, we surely do have reason for hope.

37

TANIA WILLARD

Liberating Art

Neskonlith Indian Reserve, BC

Art is at once personal and political and speaks to the beholder, always attempting to engage, persuade, move, or entertain. When the artist creates a new context for communication, uses different tools and mediums, engaging the beholder, making them part of an ongoing, interactive conversation, this is artistic innovation. And this describes Tania Willard, a multimedia artist of Secwepemc and settler heritage who is interested in the intersections between Indigenous and non-Indigenous cultures, in different art forms, and in how to communicate her messages and create spaces that invite dialogue and interaction between the artist and the viewers. Her art combines traditional stories and beliefs passed down by her family with classical art and hip-hop culture. But her innovation does not stop there. Noticing that her people did not feel comfortable visiting galleries, she decided to break down those confining walls and placed the gallery outdoors in nature itself. She invited spectators to share their thoughts and emotions, replacing a passive viewing of art with active and thoughtful participation, allowing the spectators themselves to become part of the artwork.

Tania was born in 1977 in a small town in British Columbia, now called Armstrong, and she grew up in Chase and Armstrong. When she was sixteen, she accompanied her aunt, who was selling apples at a powwow, where in addition to traditional dancing, there was also break dancing. Tania was amazed, and this inspired her interest in exploring artistic expression from different periods and in viewing art as a way to connect generations and cultures.

After high school, Tania began her studies at UBC Okanagan in Kelowna and transferred to the University of Victoria, where she completed an honours degree in fine arts in 1998. Reflecting on her education, she says her experience with the Canadian curriculum was largely negative. There was no content or reflection that would speak to Indigenous people, and she felt disconnected from her culture. On graduating, she knew she wanted to be an artist. She enjoyed learning from elders and artists like David Paul Seymour and Barbara Marchand, Dana Claxton, Lawrence Paul Yuxweluptun, and Rebecca Belmore. She met them as colleagues in art school and through her work as curator of Grunt Gallery and Kamloops Art Gallery, mounting exhibits, and creating public art projects, most notably in Vancouver and Kamloops. She also completed an MFA at UBC Okanagan in 2018.

After residing on the coast for ten years, Tania moved home and has lived for the last decade on the Neskonlith Indian Reserve, Secwepemc Territory. She returned so her children could grow up appreciating and understanding their culture and developing a close relationship to nature in a way they could not when living in the city. She particularly values the cultural activities on the reserve, including a professional outdoor theatre and a language immersion school, which is something for which she had advocated for nearly twenty-five years. She is constantly learning more about her language and the history and culture of her people. This knowledge inspires her art. For example, she recently learned a new word in Secwepemec from an elder, who told her that the word for "bead" was also the same word for a dewdrop on a leaf. She muses on how the word embodies art and nature at one and the same time. Tania reminds me that "language changes your view of place and ways of being, and there are thirty different Indigenous languages in British Columbia."

She has created an outdoor gallery on the Neskonlith Reserve. Tania describes her "lightbulb moment": she had an exhibit in Kamloops in 2009 and was disappointed that few Indigenous people came to her show. She realized that the space was challenging for Indigenous people, who were not accustomed to visiting galleries and who felt more comfortable outdoors in nature. She thus decided to bring the gallery to the land and the community. In doing so, she has made nature—and the spectators—part of the gallery. She is consciously trying to liberate art from the confines of galleries and the preconceived understanding of what actually constitutes art. She thinks about building communities and culture through land and art. The line between art and nature is often erased in her poetry, music, paintings, sculpture, and beadwork. Her work *Affirmation for Wildflowers; an Ethnobotany of Desire* is a visual demonstration of this purposeful elimination of the distance between the spectator, art, and nature. At this installation, the viewer is invited to peer into a window that contains a series of medallion-like mirrors that are framed with floral motifs and stand in front of boldly painted affirmations such as "I am the land. I am the future. I have a value." The mirrors reflect the viewers back to themselves.

When Tania describes being on the land and hearing the birds' songs, one cannot fail to think of her work, *The Liberation of the Chinook Wind.* The Chinook Wind is an animate being in the Secwepemc creation story. In Tania's art it whispers softly, reminding us that humans may think they are the centre of the world, but that they are actually the guests of nature. The exhibit includes daily poems generated from live weather reports that can be viewed at the University of Toronto Mississauga. The wind is captured in brightly coloured wind sleeves bearing the words: "water," "claim," "fight," and "thrash," which refer at once to land claims and the Pacific salmon known for fighting and thrashing. The word "Chinook" is also the hybrid of Indigenous and settler languages spoken in the Pacific Northwest.

Tania is currently working on a second master's in interdisciplinary studies and is an assistant professor at the University of British Columbia. She constantly seeks knowledge about the land, art, and traditions that include respect for nature and elders. As a curator, she studies and presents the work of others, receiving the Hnatyshyn Foundation Award for Curatorial Excellence in 2016. As an artist, she learns about the subject, the materials, and how to do what is required to complete a project. For example, for *Rule of the Tree,* an installation in the Commercial-Broadway SkyTrain station in Vancouver, she purchased all the materials and fabricated the sculpture herself. She learned about forests and trees and how to do laser etching. She learned about the Indigenous languages in the area "whose ways of relating to the land . . . are full of science, metaphor, poetry, and methods of deepening our relationship to the land."* The result is the transformation of a supporting pillar in the station into a bright red tree whose boughs reach up to the ceiling and whose base is surrounded by a large metal skirting with intricate designs and words for land and place in the Indigenous languages of the region, reminding viewers that the station is located on land that was once owned by Indigenous peoples and filled with trees.

When asked to describe herself in ten words or less on the University of British Columbia website, Tania chose: learning, growing, challenging, focusing, decolonizing, and collaborating. All active verbs. All thoughtful and indicative of her identity. She says that Indigenous knowledge challenges us all to think differently about respecting the environment and about the importance of place in human identity. The gerund form reminds me of her texts, "embraced—embracing—enclosed—enclosing," in the performance piece *Surrounded/Surrounding*, where on the one hand, viewers are invited to sit in front of a map containing words that are key to where and who we are, including family, for example. The seats are arranged in a circle around a fire pit, but at the end of the exhibit, a photo of the same map appears greatly enlarged in front of a pickup truck parked in a natural setting. We are surrounded by nature and we surround it. We embrace nature and the values of community, but we are filling it with roads and vehicles.

Tania is prompted by traditions and personal experience, and juxtaposes the traditional and modern. In her painting *Kye7e Dress*, for example, she uses graffiti techniques while borrowing from urban muralist traditions. This work features the stenciled image of a native dress projected against walls, with words declaring that the hearts of the people have not been defeated. Her work includes not only images and words but also music. She uses computers to generate the sounds and has recorded ethereal voices reminiscent of liturgical hymns. The rhythm of her music was inspired by the sound of her father gathering and chopping firewood. She admires his profound knowledge and respects him as "he enacts his sovereign rights

---

* Interview with Tanushree Pillai, *The Buzzer Blog*, February 20, 2019.

when harvesting." Tania is also a poet, and the beauty of her texts is limpid and moving.

When I ask Tania about the voice she gives to the wind, she says she "wanted the wind to write poetry." Then she had to think of a way to achieve this effect. She innovatively used a Python computer code to create the sound. As she translates her idea into art, she has a goal, but remains flexible and willing to adapt if the artwork does not happen as planned. This is akin to the work of scientists in a laboratory, where they may have a specific plan in mind to create a chemical reaction, for example, but then discover a better way to achieve that reaction. On these occasions, the process becomes more important than the objective and becomes itself an innovation and a contribution to our knowledge.

In her innovative and ironic work *Daydreamers' Tea Service*, which neatly comments on what was and could be through the contrast between settlers' tea-drinking traditions and native herbs gathered on a land that once was Indigenous territory, Tania proposed a daydreaming tea blend to be harvested and brewed as part of the *Dreaming the Land Residency* program. The Open Space gallery terms this "an affirming of plant relations and an assertion of anti-capitalist time mismanagement through intentional disruption of daydreaming expressed through Indigenous plant knowledges and the making of tea." In *Gut Instincts*, the installation is a striking tableau of a shimmering basket that also resembles a spaceship. Tania says that the artwork gives a sense of temporality, with the past and the future collapsing into each other.

Each work Tania creates is fresh and innovative. Each one crosses boundaries and engages the viewer in thought. Tania's grandfather said that he lived in "two worlds," referring to the Indigenous and Western cultures. Tania brings them together in a brilliant and innovative fashion, preserving the past and offering us all the idea of a future that will enable us to see, hear, appreciate, and value the meaning of language and art.

38

M. G. VASSANJI

The Art of Fusing Memory and Imagination

Toronto, ON

Twice winner of the Giller Prize, and recipient of the Governor General's Literary Award for nonfiction, the Harbourfront Festival Prize, the Commonwealth Writers' Prize for Best First Book, the Bressani Literary Prize, and the Canada Council for the Arts Molson Prize, Vassanji is an extraordinarily gifted writer whose work exists in a unique space of his own creation: between history and memory, myth and contemporary reality, personal and public narration, fiction and nonfiction. Vassanji creatively mixes knowledge of history with his own personal memories to produce a historical narrative. His experience and understanding of emigration and immigration, of leaving home and creating a new home, of never forgetting his former home, but knowing it has changed in his absence, all make his work particularly meaningful and poignant.

Vassanji's numerous written works, including nine novels, two collections of short stories, a travel memoir on India and another on East Africa, and works dedicated to other authors, clearly demonstrate creativity, but his story reveals a second thread of innovation. In addition to composing his own marvelous texts in which his memories come alive, he has created the space and the means for others to share their own stories, empowering these voices and enriching us all through his quietly heroic efforts.

Born in Nairobi, Kenya, in 1950, Vassanji grew up in Dar es Salaam, Tanganyika (now Tanzania), where he was raised along with five siblings and a number of cousins by his mother. Vassanji says that life was not easy for his widowed mother, who ran a store to make ends meet, but he only recalls the good times and his family's close relationship in a crowded home. He helped out in his mother's store, doing small jobs and encouraging passersby to come in to shop. His mother encouraged all the children to study, and they aspired, as was the norm, to become engineers, doctors, or lawyers. Vassanji loved reading, and every two days he would go to the library and bring home novels, which he enjoyed, perhaps planting the early seeds for his literary career.

Most of his classmates went to the universities in Tanzania, Kenya, or Uganda, and Vassanji was accepted to study in Nairobi, Kenya. However, he was encouraged by both an older student and a former math teacher to study in the US. Thus, Vassanji and a longtime classmate applied to MIT, where they were both accepted. Vassanji left home, thinking he would soon return to find his family and friends and a country once called Tanganyika. However, due to political and cultural upheavals, most of his family and friends have left what is now Tanzania and have scattered around the world. Vassanji notes that his entire life has been dedicated to creating story by story the street where he grew up. He muses, "Novels give shape to memories and writing offers coherence in understanding where we came from and what we went through."

Upon arriving in Boston in August 1970, Vassanji made his way to Harvard Square, where he planned to spend the night in a YMCA recovering from jet lag. However, he had little sleep, as the square was filled with police firing tear gas and students throwing rocks. Vassanji had no idea what was happening and was understandably alarmed. He would later learn it was a protest against the Vietnam War. Following this eventful introduction to America, Vassanji was welcomed by a host family as part of a program called "Experiment in International Living." He recalls with fondness "the generosity of Americans who took a stranger into their midst."

In 1974, Vassanji completed his BSc degree in physics, with the intention of returning home to teach. This was not possible due to the violent upheaval in his former country, so he decided to continue his studies in the US. His undergraduate program had included a requirement in the humanities, which he filled by taking courses in history, literature, and writing; the latter required that he keep a journal, a habit he continues to this day. He considers this broad exposure to literature and history partially influenced his determination to write.

For his PhD, he had to choose between engineering and applied science at Yale or nuclear physics at the University of Pennsylvania (Penn). He chose the latter, but when he graduated in 1978, jobs in nuclear physics were few and far between. He ended up in Chalk River, Ontario, following the same trajectory as Nobel laureate Arthur B. McDonald, who also began his career there. The accelerator and nuclear power facility at Chalk River had an important experimental section where research was government sponsored. However, the lab was underfunded, and Vassanji's colleagues were all leaving for other places. Vassanji, who had been on a two-year postdoctoral fellowship, went in 1980 to the University of Toronto, where he would work for nine years as a lecturer in physics and a research associate in work on which he had based his thesis. However, from the time he left Chalk River, Vassanji knew he wanted to write, to re-create the place where he grew up and a place he might never again see.

In 1981, Vassanji, his wife, Nurjehan Aziz, and two friends from Dar es Salaam met in New York and decided to create a literary magazine. Initially it was called the *Toronto South Asian Review*, but was later renamed the *Toronto Review of Contemporary Writing Abroad.* The magazine was published regularly three times a year, which adds up to six thousand pages over twenty years. It included works by authors from Southeast Asia, Africa, the Caribbean, China, and by Indigenous Canadian authors who were overlooked by mainstream publishers. Many of the works were inspired by the experience of immigration, the loss of one's home, and the discovery of a new place.

Why did Vassanji undertake this project with his wife and friends? Vassanji responds thoughtfully that it is in the Ismaili tradition to observe when something is missing—schools, hospitals, dispensaries, museums, for example—and to step up and take action. Creating a venue where authors of different cultures and races could be published was simply something that needed to be done. Canada was a country that boasted of its multiculturalism and yet the mainstream looked away from the works of authors from

Africa, South and East Asia, the Caribbean, and by Indigenous peoples. Every year Canada welcomes immigrants and therefore "has changed, is changing, and will continue to change." The magazine was intended to showcase this change and be a step toward letting those immigrant voices be heard.

Vassanji recognizes that this project came about as a result of enormous personal effort. Imagine, he says, a publication founded by two couples and run essentially by two people. His wife, Nurjehan, had a master's degree in chemistry and his two friends were a lawyer and a professor of physics—none of them knew a thing about publishing. They wrote personal letters to potential authors and "it was like a drop of water had fallen on a desert." These writers had stories to share. In the early days, Vassanji would edit the texts, then bring them to a local typesetter, return home with the pages, create the layout and cut and paste the pages, bring them to the printer, and then mail out the printed journals. When they first started publishing the review, he went to the printer at the University of Toronto Press and showed them a copy of the *Kenyon Review* and said they wanted something of similar quality. Vassanji and his friends covered the expenses for the first three issues from their pockets, but they later received grants from the Canada Council for the Arts and the Ontario Arts Council. Later still they were able to pay the writers nominal fees.

Eventually the literary magazine evolved into a publishing company and in 2015 changed its name to Mawenzi House Publishers, specializing in "writing that reflects the diversity of our rapidly globalizing world." Mawenzi is the name of the second peak of Kilimanjaro, which perhaps symbolizes the publishers' lofty goals. Over 150 books have thus far been published by Mawenzi House Publishers. Vassanji reflects that this project has meant a great deal to authors who were, like himself, immigrants. Vassanji is like an archaeologist who helps others who have lost their place and their home rediscover their history.

In 2011, thirty years after starting the literary magazine, to fill another need in Canada's writing culture, Vassanji and his wife started looking for donations and put together the Canadian South Asian Literary Festival. Vassanji saw the festival as a challenge to the Toronto International Festival of Authors, which was "essentially all white." The new festival would include three days of readings, a dance performance, and a dinner. It was so successful that the event has been held every two years since and has included writers from the Caribbean, Africa, and South and East Asia, as well as local Canadian authors. Now, after forty years, Vassanji and his wife look forward to finding successors who will take, as they do, great satisfaction in seeing the desert continue to bloom in Canada.

Vassanji has given us fragments of his history and more are sure to follow. Having written about the place where he grew up and his memories of his mother and grandparents, he embroiders on his experiences and memories, imagining other narratives about people who may, or may not, have existed. He recently completed a novel about a physicist who may, or may not, be inspired by Vassanji's work in that field. He is also writing a collection of essays. These days he travels to Portugal each year for two or three months. He isolates himself completely for six weeks and says, "Loneliness is a nice feeling." He is then joined by his wife and they travel together and he reads, listens to music, and continues to write. He could have pursued other occupations that would have made more money and would have given him stability, but he says, "I would not have survived. Stories come out of your experience, and they are what you are. If you are a writer, you are a writer." He describes his first trip back to Africa in his 2014 travel memoir, *And Home Was Kariakoo: A Memoir of East Africa*, as an overwhelming experience: "I saw the land where I grew up, kicked the sand, walked to school, played ball. The memories do not go away even if you no longer know the people." Vassanji's memoir is historical, descriptive, and contemplative. In it he rediscovers a place he once knew, whereas in his novels he creates a fictional place coloured with images from the fading photos in the album of his mind.

As a publisher, Vassanji has given a voice to dozens of talented and otherwise underrepresented writers and enabled them to share their memories and stories of former homes. If a country is a cloth woven from the many narratives of the people who live there, Vassanji has offered us a vibrant tapestry that includes his own work and the work of many others who now call Canada home. May he long continue weaving narratives and empowering others to tell their tales, for these stories constitute our national treasure.

39

S. GORDON HARWOOD

When the Baton Becomes a Brush, Painting Music

Ottawa, ON

Born in Ailsa Craig, Ontario, S. Gordon once travelled to his hometown's namesake, an uninhabited island in the Firth of Clyde, Scotland. Legends would have it that Ailsa Craig was used as a stepping-stone by a giant crossing the Irish Sea on foot. In his art, S. Gordon is unique and his work is like a stepping-stone between art and music, creation and performance.

S. Gordon's paintings are highly prized in France as well as in Canada. His work combines bold colour with an evocative frame of reference. He completely immerses himself in music for days and weeks on end, and innovatively translates the musical notes into brushstrokes. His paintings capture the music and the mood it inspires. For example, S. Gordon has created a suite of works dedicated to a number of jazz artists. As you view the works, you can well imagine the rhythms of the music and the sounds of the different instruments. S. Gordon's process of creation may include paint, etching, total abstraction, or a combination of images and abstraction. Each work reaches out to the viewer and demands a response. He is at once inspired and inspiring.

Born in 1955, S. Gordon had a childhood that was an introduction to self-sufficiency. His parents encouraged their children to work hard and to do their best. S. Gordon had many jobs, including cutting lawns with a push mower and working as a cleaner in a turkey barn, which convinced him that "there must be better jobs out there!" While S. Gordon was encouraged to choose his own path, he was given a bit of "nudging" by his father. Although he himself had little education, S. Gordon's father had an ear for music and played the piano and fiddle, which he had learned from his own mother, who used to play the piano with a gospel band on the radio every Saturday evening. His father appreciated the power of music to delight and inspire, and so one of his "nudges" was to purchase guitars for his children. While this did not prove successful when S. Gordon was nine years old, it had a delayed effect: S. Gordon bought another guitar with his first paycheque when he was eighteen, and it signalled the beginning of his lifelong love for music, which is a fundamental part of his creative process.

S. Gordon was also interested in computers and data processing. He was fascinated by keypunch machines and figured out how to program them. In 1973, he went to Fanshawe College in London, Ontario, graduating as a software developer. After working in finance on payroll systems for nine years, he was accepted directly in the finance stream of the MBA program at the University of Western Ontario (now Western) in London. Upon graduating in 1987, he was recruited to lead a project management program, which he had designed to help people prepare for exams. He enjoyed both creating the program and teaching in it and "nudging," in his father's tradition, people along the path to success.

In 2000, S. Gordon was offered work at an aerospace company that made missile guidance systems. But the Gulf War broke out, and S. Gordon decided to take a break from the industry and signed up for a painting holiday in France. In June 2006, he found himself sitting in the middle of a lavender field with his easel. A gentle breeze wafted the scent of the flowers, and S. Gordon sat back and looked at his canvas and realized that painting was something he had to do. The experience of that day signaled a new direction in his life, and two years later, he decided, with the support of his wife, to pursue a new career as an artist. He felt a bit as if he had "made a left-hand turn without signalling." It was certainly a financial risk at first, but he knew he had found his true passion and discovered his talent. He knew he was on the "right path."

He studied the works of great artists over the centuries and began painting landscapes and portraits. Then he discovered the works of painter and sculptor Jean-Paul Riopelle at the National Gallery of Canada and that "changed everything." He made another left-hand turn on seeing Riopelle's 1954 triptych, *Hommage aux Nymphéas—Pavane* (*Tribute to the Water Lilies—Pavane*). He felt its emotional power. He notes that "a pretty picture is just pretty. You walk away and never remember it. But look at Van Gogh's last painting of the crows. It exudes his personal suffering." S. Gordon visited the spot where Van Gogh did that painting, *Wheatfield with Crows*, as well as the room where he died and his grave. He now felt a strong link between emotion and visual art, and he wanted to capture that connection in his own painting. He found that same close relationship with music, especially jazz. He thinks about the circumstances in which the music was created and about the moment when jazz comes together. The musicians make the instruments talk to each other. When you listen to jazz you hear the instruments being played together, but then there is also a moment when one instrument takes the lead and invents a new melody or shifts the rhythm, telling a personal story. Then another instrument will take the lead and respond, embroidering on the narrative. It is as if the instruments (and their players) were having a conversation and sharing it with us. S. Gordon reproduces that dialogue visually. His hand and the brush become one. He becomes totally immersed in the music. He mentions the drum solo in "Take Five," in which the percussion actually carries the melody. He says, "This is true genius, and it is at such moments that people become egoless." These are moments that S. Gordon makes eternal on his canvas.

S. Gordon quotes acclaimed trumpeter and composer Miles Davis, "A painting is music you can see, and music is a painting you can hear." This captures S. Gordon's new approach to art, as he began to paint with his emotions and with music.

He then took one step further to include his audience in his work and to create in public while listening together to music.

In 2017, a former colleague from S. Gordon's days in finance had just returned to Canada from abroad and by chance reached out and invited him to visit Archambault Institution, a maximum-security prison in Sainte-Anne-des-Plaines, north of Montréal. This colleague had created a foundation, Looking at the Stars, to bring music to prisons and long-term care centres. S. Gordon agreed to attend a concert performed by the highly acclaimed Russian-Lithuanian pianist Lukas Geniušas, and on a whim, S. Gordon asked if he could bring his paints. This would turn out to be another left-hand turn.

The concert was scheduled to begin at 7 p.m. and last one hour. However, since the Montréal Canadiens were playing the New York Rangers in the NHL playoffs that night, the concert was moved to 6:30 p.m. and the prisoners were told they could leave at 7 p.m.

At the concert, S. Gordon stood facing his canvas, with his back to the prisoners and Lukas, who started to play Beethoven's sonatas. S. Gordon was mesmerized. It was so beautiful, and he was totally "blown away." At 7 p.m., the gate went up so the prisoners could leave to watch hockey. None left. Indeed, others joined them. Lukas looked at S. Gordon, who had not even picked up his brush, and said, "Paint!" Inspired by the music, S. Gordon did what he considers was "possibly the best piece [he] ever created," and it is now in the cafeteria of the prison. He only finished painting at 8 p.m.—well into the hockey game's second period—but the prisoners had not left. Instead they came up and asked him and Lukas to autograph their programs. Perhaps unsurprisingly, S. Gordon now serves as artist-in-residence at Looking at the Stars.

Since that evening at Archambault, S. Gordon has often repeated the same experience with volunteer musicians from the Toronto Symphony Orchestra. He is profoundly moved by what music can do for the soul. Seeing men weep in prison on hearing Chopin's Études, he was touched by the tenderness that the joyful music inspired. When asked why he chooses to paint in prisons, S. Gordon says that "we send people to prison *as* punishment, but not *for* punishment." He hopes to provide inspiration, a space for contemplation, and a sense of peace through his works. S. Gordon's innovation resides in his sensitivity and his humanity, which let him deeply experience the feelings of others and communicate the beauty of music combined with art.

Not many artists share their works with prisoners and very few accept to share their creative process, which is normally private. The public rarely sees the parts that are painted over, the mistakes. While we can go to galleries and museums to view paintings, rarely are we invited into a studio to see an artist actually create a painting. Art on demand is a concept that strikes fear in the heart of most artists. They worry what might happen if inspiration were elusive that day, and it can also be stressful to know they have one hour to produce an original masterpiece. But this is the task S. Gordon sets himself. He creates powerful works of art while immersed in an atmosphere highly charged with emotion and filled with music. S. Gordon's artistic innovation comes from capturing the moment in which music, emotion, and visual art become one.

S. Gordon offers the following advice: "Take a few risks. Life is in risks. Make some mistakes and try again. Do not be afraid to break the rules. Make a left-hand turn or two." Remember that he made his way through "some of the worst jobs in the world" and sought fulfillment in a business career. But he ultimately found art, which has given him joy, depth, and meaning—and he has chosen to share his genius not only with the general public but also with some of those who most need meaning and joy.

40

TINA FEDESKI AND MARGARET MARIA TOBOLOWKSA

Building Community with Art

Ottawa, ON

This is the story of two accomplished musicians, one a flautist and the other a cellist, both born in Toronto. They would only meet years later in Ottawa, Ontario, where they, along with Gary McMillen, would create OrKidstra, a not-for-profit organization that provides access to instruments and musical training for youth in underserved communities. Its goal is to build life skills and instill confidence, kindness, caring, and a sense of purpose, and its students come from sixty-two different linguistic and cultural backgrounds. One of their graduates, who was born in a refugee camp, recently said, "I am who I am today because of OrKidstra. Being part of OrKidstra taught me to be a better person, to accomplish what I set out to do, to practice with care and to do what I do with love." This is a story of courage and talent, perseverance, and passion, and dedication to innovation in art, community service, and teaching.

Tina was born in Canada in 1962, and her parents emigrated back to England when she was four. Her parents, who loved music but did not have the means to learn an instrument when they were younger, gave her a recorder and piano lessons. When she was eight, Tina was enrolled in a new school that offered music lessons and was asked if she would like to play an instrument. Remembering that she had once heard a wonderful flute solo at a concert, she said without hesitation she would like to play the flute. Tina describes it as "love at first sight." She would take the flute from her bag during spelling class so she could look down at what she considered the most beautiful object in the world.

Tina later played in the Bedfordshire Youth Symphony Orchestra, which represented England on world tours, and went to the Guildhall School of Music & Drama, where she studied with Britain's top musicians and received an outstanding education in music. If she had to do it again, she thinks she would seek a broader education in a university before specializing in one instrument. She is keenly aware of the extraordinary privilege she was given and recalls being disturbed by the lack of racial, gender, and economic diversity in the composition of most orchestras and the financial barriers that prevented people from attending concerts. She spoke with people in the subway and found that only a few had been able to enjoy concerts, responding to her inquiry with "Not for the likes of me, luv!" This inspired her to want to make orchestras more accessible and appealing to a broader audience. She designed a line of clothing for orchestra players that replaced the somber black tie/black dress uniform with more imaginative and colourful clothing and was invited to do a fashion show with the London Philharmonic. She was about to launch a business that showed her creative spirit when she was invited to Tenerife, where she dedicated her talents to playing.

Margaret Maria was born in 1976. She was the child of immigrant parents, who listened to Polish folk music and gave her piano lessons. She recalls that there was a concert at her future middle school, where she saw the cello played by a girl using vibrato, a technique that slightly raises and lowers pitch to create a sense of warmth and emotion. Margaret Maria was "completely taken and fell in love with the cello." When she was thirteen, she had her first cello lesson and was sent to an amateur orchestra. She initially sat in the back because she could not execute all the notes as rapidly as required, but by her second year, her talent and ability were recognized, and she was awarded a prominent seat next to the first cellist.

When she was fourteen, her parents managed to buy her a $5,000 cello;

and with her own earnings from part-time jobs, she paid $1,500 for the case and the bow. She always knew that she wanted to be a musician and was so determined that her parents supported her without really understanding what she was doing. She applied and was admitted exceptionally to a bachelor's degree in music performance at the University of Toronto. She was only sixteen and never completed high school. If she had remained in high school she would have had to pay for private lessons, which she could not afford. At the university, however, she had classes with the first cellist of the Toronto Symphony Orchestra as part of her regular course load. Margaret Maria says she "flew by the seat of my pants, but music is so amazing that you can achieve your dream even if you are not well off." Upon graduation in 1992, she went on to study for two years at the renowned Curtis Institute of Music in Philadelphia, where she earned an artist's diploma in 1994.

Margaret Maria immediately got her first professional job at the Vancouver Symphony Orchestra, where she remained for four and a half years before going on to the National Arts Centre Orchestra in Ottawa. While she enjoyed playing, she always felt there must be something more in life beyond beautiful music. There must be something she could do to help fix some of the wrongs, to solve a few of the problems in the world. After a traumatic life experience, she began to "hear her own music and to compose." She took a sabbatical from the NAC Orchestra and started making her own arrangements, compositions, and recordings. She credits the path on which her life took her with the fact that she was "now writing as if there were no tomorrow."

For her part, Tina graduated from the Guildhall School in 1985 and began her professional career as principal flautist at the Tenerife Symphony Orchestra in 1986, where she played, toured, and recorded for four years. She was "in paradise," but wanted more, to be part of something bigger. She did not know what it might be, but she felt isolated in the Canary Islands and knew, in her heart, that there was something else she had to do. So, when the Banff Centre for Arts and Creativity in Alberta offered her a six-month fellowship in 1990, she courageously packed up and left Tenerife. After her fellowship, she joined her sister in Ottawa. She busked on Sparks Street, freelanced with major orchestras and festivals, and taught. During this time, she set out to create and sell practical and yet beautiful items including, for example, soap dispensers for hotels that solved problems such as environmental degradation. She also met her future husband, Gary McMillen, an electrical engineer who had previously been a rock drummer and is now a passionate cellist. Together they created an enterprise called the Leading Note, a print music store, and made a solemn promise to "give back to the community" by breaking down the barriers that prevent people of different races, genders, cultures, and economic backgrounds from being musicians.

Tina was later inspired by a splendid recording of Beethoven's Fifth Symphony, in which the players were "brilliant, passionate, and inspired" members of the Simón Bolívar Symphony Orchestra of Venezuela. Tina learned about the orchestra and El Sistema, a publicly financed music education program. Its motto was "Music for Social Change," and the program provided instruments and instruction to some seven hundred thousand young musicians in four hundred centres under the leadership of José Antonio Abreu, its acclaimed founder. Gary ordered a DVD of a documentary about El Sistema as a birthday present for Tina. It arrived a day early, but neither could wait—they watched it right away and both ended up in tears. They said, "That is it! This is the way we will give back! We are going to do something like El Sistema here in Canada." It took a few years to find the funds and to create the administrative structure as well as a way to replicate the philosophy in Canada. Tina went to Venezuela to research El Sistema in 2007. In addition, they needed superb teachers with a special pedagogical philosophy.

Tina remembers a magical performance of the Martinu Trio in 2001 when she played with Margaret Maria in a concert at the University of Ottawa. Margaret Maria recalls that Tina and Gary often came backstage after shows at the National Arts Centre to congratulate performers. Everyone knew them, and their shop was the "hub for all classical music and musicians in Ottawa." When Margaret Maria stopped by the shop that January, Tina gave her the DVD to watch. Margaret Maria agreed with the philosophy of El Sistema and was immediately "sold," joining Tina and Gary in the adventure and becoming the star teacher they needed.

Margaret Maria had been experimenting with spontaneous music creation, breaking down musical structures and thinking differently. This exploration led her to develop a new way to guide children's creativity, including their ability to explore and improvise. She began encouraging students at a young age to compose by creating sound paintings. Instead of asking them to paint a picture of something, she asks them to listen to the music and paint what they hear. She does not confine the children to specific structures or theories.

Tina and Margaret Maria adopted a "child-centered approach" and tried to put the ball firmly in the court of children who would not be taught by rote learning, but be encouraged to ask questions. They also included peer-to-peer teaching with students showing one another what they had learned. Tina and Margaret Maria initially struggled to balance excellence with diversity and inclusion, passion with precision, and the dual goals of being an outstanding musician with being a good and kind human being. But they persevered and succeeded through their innovative teaching methods and the warm, welcoming atmosphere they created. They ask students to make a commitment to mentor future students in the program, creating a sense of continuity and family. Tina stresses the fact that in an atmosphere of trust, young children are able to belong in a way that transcends differences. They are given a special opportunity and are made to feel part of a family where they learn together and share a love of music.

Tina recalls the first classes each year when the youngest students are introduced to instruments and allowed to experiment with them. They "honk, blow, and screech" until suddenly their faces break out into smiles as they start to make music. Sometimes she thinks her heart will burst when she sees children putting their entire soul into playing. Margaret Maria muses how fortunate she was that "the cello found me," while Tina echoes the thought and remembers how she chose the flute . . . or perhaps it was the flute that chose her! With OrKidstra, they are thrilled to see children finding such joy in music.

OrKidstra works because of its dedicated staff, teaching artists, volunteers, and ambassadors, and the community that come together to encourage and support these young musicians. OrKidstra in turn serves the community, delighting attendees at charitable events with their enthusiastic performances in a wide range of musical styles. Tina would like all Canadians to discover joy by breaking down barriers, having a sense of teamwork, and belonging to an organization like an orchestra, which is a large family.

Margaret Maria describes an encounter with a very young lad when she went to play at a local elementary school. He came up to her after the concert and said, "I want to play the sello" (he could not yet pronounce "cello"). He was simply drawn to the instrument, and Margaret Maria remembered falling in love with the cello herself. She recalls her childhood dream of wanting to play in an orchestra. Today she surpassed that dream, and not only creates her own music, but also sets young musicians on a path to achieve the same. But it is not only about musical success—it is about community and the hugs parents and students share after a concert. Margaret Maria says that even if you help only one child, you are contributing to a kinder, more beautiful place. We can all mentor and be more inclusive. We must learn to play every note with love, and respectfully listen and collaborate with one another—actions she and Tina endeavour to teach their students.

In 2012, a virtual, live, and unrehearsed concert occurred at Carleton University in Ottawa. Players from the Simón Bolívar String Quartet, who performed virtually from Venezuela, were joined in real time and in person by members of the National Arts Centre Orchestra, the Carleton University Orchestra, and OrKidstra. They all performed together, at first tentatively and then with growing confidence and obvious pleasure and passion. The performance showed what children could accomplish when afforded the opportunity, instruments, and coaches. It also confirmed that music is indeed an international language that transcends all barriers of race, culture, and economic status. It was a small but powerful demonstration of what is possible when young people are given the opportunity to learn and play with extraordinary artists who generously share their artistry. It was both innovative and inspiring—like OrKidstra and its founders!

PART SEVEN

Engineering Innovation

From lithium batteries to radar systems that map the outer atmosphere, from antimicrobial surfaces to microscopes remotely controlled via Twitter (now X), these innovations all contribute to improving our lives by taking on a challenge. They use scientific reasoning, employ technology in new ways, or, when no appropriate technology exists, they simply set about to create new technologies.

Each of these engineers is unquestionably an international expert in their field and has a special talent, passion, and vision. They share deep knowledge of their areas of research and are indefatigable in their search for solutions. For Jeff Dahn it is not simply a matter of creating a better, lighter, longer-lasting battery, it is a question of the materials used and their impact on the environment, their availability, cost, and efficiency. For Kathryn McWilliams it is not only a matter of tracking the speed of particles in the Earth's near-space, but a quest to monitor space weather conditions to predict disturbances that affect air travel and can cause major power outages around the world. To undertake this project with five radars located in northern regions of Canada, each scanning four thousand square kilometres every minute, twenty-four hours a day, 365 days a year, meant innovatively updating the equipment. The brilliant design created and built by Kathryn and her team is now being adopted in other countries around the world.

Anna Kazantseva's novel work began, on the other hand, with the technology that enabled the transcription of language and the ability to synthesize texts. She applied the technology to a real problem, the disappearance of Indigenous languages in Canada. In partnership with native speakers, she is providing user-friendly tools that enable the study and preservation of language.

Leyla Soleymani's research begins with the problem to be solved, and she uses the materials at hand to do this. However, she does not stop there. She continues her work until the solution is marketable and serves the population. Whether it is antimicrobial films to protect against the spread of viruses and bacteria, portable devices for people with diabetes, or testing units to enable family physicians from their offices to diagnose illnesses and efficiently and rapidly prescribe medicines needed by patients, Leyla is prepared to meet the next challenge. Priti Wanjara looks at problems that arise in manufacturing and solves them effectively and inexpensively, saving companies and consumers much time and expense. While doing so, she invents new ways to fuse metals and create stronger, lighter, and more flexible materials for aircraft, for example. Teodor Veres works with polymers and creates the nanomaterials that can enable new technologies with a myriad of applications from communications devices to space stations. Andrew Pelling deliberately seeks innovation in a broad array of areas, in zones where nobody has looked before and in those which have been overlooked or dismissed as impossible or improbable. He then uses scientific methods and logic to create cutting-edge technologies that have wide and varied applications. While some innovators are on a quest for solutions to a problem or for ways to enable the solution, Andrew asks the questions that will themselves spark new ideas and ways of looking at challenges.

41

KATHRYN MCWILLIAMS

A Star among Stars

Saskatoon, SK

In 1993, representatives of Canada, France, and the US met in Saskatoon, Saskatchewan, to turn on the first SuperDARN radar in a network that today includes over thirty-five high-frequency radars located in both the northern and southern hemispheres and operated by ten countries. Canada has five locations, and each radar scans more than four thousand square kilometres every minute, twenty-four hours a day, 365 days a year. Each location is remote and possibly even colder than the one before: Saskatoon, Prince George, Rankin Inlet, Inuvik, and Clyde River.

SuperDARN stands for Super Dual Auroral Radar Network. These radars operate a bit like highway speed radars, sending a signal that meets an object and bouncing back like an echo, capturing the speed of a vehicle. SuperDARN's radars beam up into the sky, where they collide with electrically charged particles and scatter the signal back, allowing the researchers to determine the particles' speed.

In this way, researchers can monitor weather conditions in the near-Earth space environment, or the magnetosphere. Disturbances in space produce the beautiful northern lights, but they can also affect power grids, satellites, radio communications, and cause airplane and space station radiation hazards. The danger is not simply speculative. On March 13, 1989, a geomagnetic storm caused power blackouts along the Eastern Seaboard of Canada and the US. In August of the same year, another storm caused the Toronto Stock Exchange to close and shut down computing systems across the country. In 1994, in a single day, a solar storm knocked out Anik satellites worth $100 million, disrupting television signals, telephones, and computers. And in 2012, a large amount of plasma and magnetic field were ejected from the sun's corona. This coronal mass ejection (CME) narrowly missed the Earth by just nine days.

These storms occur each time there is a big explosion on the sun, ejecting energy and gases into space that may even reach the Earth. Scientists would like to be able to predict the path of these magnetic storms and their interaction with the magnetosphere. The data they collect will assist in understanding the causes and timing of these storms, protecting human lives. This is a compelling time to research this topic, as the sun is in a cycle of increasing activity until 2025. Canada is certainly the right place to do this work, as it has the largest landmass under the auroral oval.

SuperDARN is led by a researcher from Saskatchewan who also heads the international SuperDARN collaboration and chairs its executive council: Dr. Kathryn McWilliams, who is also the first Canadian to be recognized by the Royal Astronomical Society with an honorary fellowship for her "unquestioned international expertise in the dynamics of field-aligned currents that link solar wind, magnetosphere and ionosphere." When she learned of her honorary fellowship, she carried on with her day and completely forgot to share the good news with others—including her mother! On the university website in the "About Me" space, where professors list their achievements and interesting facts about themselves, Kathryn includes only a link to a video of a spectacular aurora over Saskatoon. It is said that a picture is worth a thousand words, and this one speaks of Kathryn's work, of the beauty of the auroral displays, and of the fact that she always places others and her research ahead of herself. She is a brilliant scientist, but also an extremely humble and modest person.

Kathryn and her family moved from Calgary to Saskatchewan when she was an infant. Her father, the first member of the family to graduate from university, became a civil servant for the provincial government and followed his career path across the province, taking his family to Swift Current,

Moose Jaw, and Regina. As a result, Kathryn was close to her family, and she and her siblings valued their education. She was always interested in math and logic, and was inspired by her science teachers, most of whom, except for a "great female chemistry teacher," were male. She enrolled in a BSc program at the University of Saskatchewan in 1990, where she continued to be fascinated by science, especially physics and engineering.

Notwithstanding the inspiration of her teachers and her decision to do a science degree at the University of Saskatchewan, one might well ask how she ended up at SuperDARN. She responds that she seemed to accidentally be in the right place when opportunities arose. The first of these occurrences may have been her meeting with engineering physicist George Sofko at the University of Saskatchewan. One of the founding members of SuperDARN, Sofko arrived in Saskatoon on a Woodrow Wilson fellowship in 1960. For his PhD thesis, he built an instrument that measured the polarization of auroral signals, and the study of the aurora borealis has been his life's work ever since. The University of Saskatchewan's campus news reported that he had a eureka moment while playing solitaire—he realized that geometry was key to understanding the causes of magnetic storms. In 1992, Sofko gave Kathryn, then a second-year university student, a summer job—building the Saskatoon site for SuperDARN. On completion of the project that summer, Kathryn and her fellow students left handprints in the cement. The handprints are still there, inspiring young researchers at the university and symbolizing the foundation on which Kathryn's career in science and innovation began.

At SuperDARN, Kathryn discovered her passion for mixing classical physics and practical system development. During the course of her studies she received two International Association for the Exchange of Students for Technical Experience (IAESTE) internships, which enabled her to go to England. In 1996, she made short-term visits to Imperial College in London, where she completed her MSc in 1997, and went on to the British Antarctic Survey, which operates a SuperDARN facility at Halley Research Station in Antarctica. In 1998, she received a Commonwealth Scholarship, which took her to the University of Leicester, where she earned her PhD in physics in 2001, working with the Radio and Space Plasma Physics Group that operated the CUTLASS radars in Fennoscandia, the region that includes the Scandinavian Peninsula, Finland, Karelia, and the Kola Peninsula. Her thesis examined solar winds and magnetism in the ionosphere, which is a layer of the Earth's atmosphere about thirty to fifty miles above the Earth's surface, with a high concentration of ions and free electrons that can bend and scatter radio waves.

In 2002, Kathryn returned to the University of Saskatchewan as a postdoctoral fellow and became the first woman to receive tenure in the Department of Physics and Engineering Physics, where she later became undergraduate program chair. Kathryn studies not only the magnetic fluctuations in the aurora seen from the ground but also the particles detected in the upper atmosphere (the magnetosphere). Her work, which includes consideration of the solar winds and physical locations, is at the crossroads of physics and engineering. Her most notable innovation is the new radar system, Borealis. She realized that the radars at SuperDARN had stood up well in a harsh climate, but they were aging, and replacement parts were no longer available. Computer systems could no longer support their software and funding was an issue. Kathryn wanted to design a system that would last another thirty years and be scalable, user-friendly, and as adaptable as possible. She was also determined to upgrade the system using readily available parts, which was a creative and resourceful plan. Even more innovative was the concept of changing the way the radars processed the data from the antennas. Previously, the signals ran through cables to combine the data from each separate antenna and could only be processed in one way. The new Borealis system allows researchers to sample the data from each antenna and then process it in many ways. Now the scientists can run full field-of-view snapshots with sixteen times as much data every 3.5 seconds. The new Borealis radars can also run at more than one frequency simultaneously, which was previously not possible. These radars will provide new capabilities, including a full-field snapshot, reducing ambiguity, moving in time and space, and isolating spatial movement. It will be, Kathryn says, a matter of "figuring out the geometry" and developing signal processing to transmit and receive at two different sites. The new Borealis system is now being adopted around the world.

Kathryn credits her engineering team, graduate students, and postdocs.

"They are amazing," she says. "They accomplish what most people would not even try."

Kathryn continuously strives to offer her students experiences that will inspire them, as her own work experience as a student at SuperDARN did for her. She organizes special programs such as CaNoRock (the Canada-Norway Student Sounding Rocket) program, which challenges students to build and launch rockets to test and assess the systems they've created and the data they measure.

She encourages sharing and mutual respect in a positive and creative workplace.

"People," she says, "who bring the best get the best results."

She constantly creates the context for new ways of thinking and solving problems. She says that one needs to push ideas and mechanics to the limit, and then one will find solutions. Infrastructure and software should not define our ability to find solutions. They should motivate us to go beyond described and prescribed capacity, to raise our sights and redefine the problem.

When asked what the future holds, Kathryn says that every country has different policies about open data and unique requirements and guidelines. She worries about how political forces have an impact on people's attitudes about misinformation and the polarization of views. She would like to be part of a worldwide effort to link the observations made at the SuperDARN locations with those made by the satellites orbiting the Earth to unlock the mysteries, the things yet unknown, to understand the sun and perhaps one day to be able to predict solar explosions and the resulting weather conditions in the Earth's atmosphere, saving lives and securing electronically based systems. She would like to have the opportunity to help everyone understand how science works.

Observing Kathryn's contributions to date and her close work with researchers in ten other countries, it is evident that her strategic thinking and ability to bring ideas and people together will contribute to the science that will solve the world's problems.

42

JEFF DAHN

Batteries for the Future

Halifax, NS

Fellow of the Royal Society of Canada and winner of the Herzberg Medal, the Medal for Excellence in Teaching Undergraduate Physics from the Canadian Association of Physicists, the Rio Tinto Alcan Award from the Chemical Institute of Canada, the ECS Battery Division Technology Award, the Yeager Award for the Battery Materials Association, and the inaugural Governor General's Innovation Award, Jeff Dahn acquired a taste for engineering and ingenuity as a youth, repairing farm equipment in the countryside of rural Nova Scotia. Born in the US, Jeff and his family relocated from Pennsylvania to Nova Scotia in 1970. Jeff recalls moving from a junior high with two thousand students in grades seven and eight alone, to a K-12 school with a total of four hundred students across all grades. In the US, only the top athletes could play on the baseball and football teams. In Canada, Jeff said, "If you could breathe there was a space for you on the team." He saw Canada as a land of opportunity where, if he followed his father's example of hard work, determination, and focus, he could achieve his goals.

The lessons his father taught him were to prove useful throughout his life and his career. His father would set tasks for Jeff, his two brothers, and sister, which might include building a chicken coop or fixing the aging tractor when it broke down. Jeff said that's how he learned basic mechanics. After high school, Jeff went to Dalhousie University in Nova Scotia and earned a bachelor of science in physics in 1978. Dr. Manfred Jerico, who worked in low-temperature physics, was Jeff's mentor and gave him a job in his lab. Jeff also recalls Dr. Gerhard Stroink, a pioneer in the field of biomagnetism, who also hired him to work in his lab. Stroink was studying the magnetic signals between the lung and the brain, and Jeff was assigned to build filters for the signals. Jeff enjoyed the interdisciplinary aspects of the two labs, and throughout his career he has combined physics, engineering, and chemistry in his projects, working on the borderline across the three fields. However, his passion was lithium batteries and they have always been central in his teaching and research.

Jeff earned a PhD in 1982 at the University of British Columbia and went directly to the National Research Council, doing research at the Brookhaven National Laboratory in Upton, New York. He worked there from 1982 to 1985 on the fundamental materials used for batteries and published many papers with his friend R. R. Haering, a professor at Simon Fraser University, who worked on lithium intercalation. Jeff remembers a pivotal moment when he was sent to another professor's lab to borrow a piece of equipment. Neatly stacked on the shelves were the offprints of all of that professor's work. The sheer number of papers was impressive. Equally remarkable was the fact that the copies of his publications had not been shared and simply remained on the shelf. Jeff reflected on so much hard work going unnoted and thought he should move to a different milieu, where he might make a bigger impact.

He sought a faster-moving and high-pressure environment and found it in Moli Energy. A young company bristling with energy, Moli had been in business for only seven years and already employed sixty passionate people with an average age of twenty-eight, working on the commercialization of rechargeable batteries.

While fully employed in industry, Jeff continued researching and publishing with a view to keeping open the door to teaching one day at a university. In 1990, he was offered a position at Simon Fraser University. He enjoyed teaching, but in 1996 concluded, together with his wife, a Vancouver native, that in view of the exponential growth of the city, Nova Scotia would be a better location to raise their children. Jeff retraced his steps back to

Dalhousie and was immediately offered a position—but one without pay. He would have to raise the funds for his salary. This obstacle did not deter Jeff, who was awarded a joint NSERC/3M Chair funded by the Natural Sciences and Engineering Research Council.

In 2000, he went to a conference in Phoenix, Arizona, and brought several graduate students along. To save money, they stayed in a "fleabag hotel six miles from the conference centre." At the conference, they listened to a talk about lithium-ion batteries, and Jeff thought the explanations just did not make sense. On the bus ride back to the hotel, he entered into a deep conversation with his postdoctoral fellow Zhonghua Lu. They started drawing diagrams on paper and were able to figure out the correct solution, which involved NMC (nickel, manganese, and cobalt). On their return to the hotel, Jeff called 3M, and the company was so impressed, it immediately filed a patent. Seven days later, a Japanese company applied for the same patent.

"Just think what would have happened if we had not stayed in that cheap hotel," Jeff muses. "Just think—if we had not been obliged to take the bus!"

Innovative ideas come from hard work, discussions with colleagues, and sometimes just a bit of serendipity combined with genius! Jeff filed twenty or thirty patents with 3M and trained many students who went to work for the company. When he won the Governor General's Award for Innovation in 2016, Jeff was asked to define innovation and responded that it was "thinking outside the box, having a good idea, and then tracking it down."

Jeff is passionate about his research and takes special pleasure in his students, who share his enthusiasm. He mentions a current student, Tina Taskovic, who is interested in what happens to lithium electrolytes when batteries die. In Jeff's lab, they seek to extend the life of batteries, to make batteries that weigh and cost less, and to find ways to contribute positively to the environment. Tina exposed the degraded material containing lithium to high temperatures, resulting in a new, more stable material that has the longest lifetime of lithium batteries ever produced. Her first paper explains the results of her experiments.

"Best of all," says Jeff, "she has discovered her life's passion."

Another student, Dr. Chris Burns, started a company called NOVONIX, with 120 employees; half of them produce synthetic graphite for lithium batteries and half do research to make high-precision batteries. The company was recognized on the Nasdaq stock exchange in February 2022. Jeff was delighted to accompany Chris to ring the bell, just as he had been proud to see Chris awarded his PhD at Dalhousie. When Chris was a student, he worked in Jeff's lab, and now Jeff is a research advisor in Chris's lab.

In 2016, when Jeff's NSERC/3M Chair ended, he read about Elon Musk's Gigafactory, where electric vehicles and the lithium batteries that power them were made on the same site. Jeff had connections on the West Coast with former students and friends who were working at 3M and Tesla. He called Kurt Kelty, now the vice president of commercialization and battery engineering at Sila Nanotechnologies Inc. (but at the time he led the battery cell team at Tesla), and asked if he thought the Dalhousie lab might be able to partner with Tesla. Kurt arranged for Jeff to meet J. B. Straubel, chief technology officer at Tesla, to make a thirty-minute pitch with no more than five slides. Jeff was very nervous, and J.B. was very late, arriving with his lunch in hand. He asked Jeff not to show his slides and not to give his pitch, but instead to show him pictures of his lab. Jeff gave a virtual tour on his laptop. J.B. decided he had to see the lab in person. Not too long after, he arranged for his plane to stop in Halifax. After touring Jeff's lab at Dalhousie, J.B. declared that he had "never seen anything like it before. The lab was innovative. Jeff and his students were fearless builders." They designed and built themselves whatever parts were needed. At the time of the visit, they were measuring the amount of gas generated in lithium-type cells. Jeff, ever the teacher, hastily calls up an image. "Imagine a beach ball with a small hole in it, submerged in a tank of water, and connected to a scale. As the gas escapes, the balloon weighs less. And voilà you have a way to measure the gas." Connor Aiken, a graduate student in the lab, built a device that does exactly that. The students ended up building twenty-four of these devices so they could simultaneously take multiple readings. This device is valuable in detecting when a gas leak occurs and determining the size of a leak.

Jeff is now thinking about the availability of resources to meet the scale of planned electrification in the future. The current amount of fossil fuels being used is mind-boggling. Vehicles are only a small part of the problem. Fossil fuels are used for many other purposes, including heating. Jeff sees the need for many more factories, and we will require 500 million tons of lithium to be able to provide sufficient electricity for vehicles and for large buildings, including manufacturing plants. There will not be sufficient supplies of lithium to build the necessary batteries. And so, the researcher known for his work on lithium batteries plans to turn his attention to sodium-ion batteries. They already exist, and China is leading the way in production. There are, of course, challenges. The density of sodium is lower than lithium, which would mean a shorter driving range per charge. Jeff is looking to invent a way to make a bigger charge. The sodium batteries are also larger than lithium batteries, and, unless the size can be reduced, drivers will have to leave their children or any other back seat occupants, including Marly the dog, home. Jeff is determined to find a way to produce these batteries in Canada and to make them smaller and more efficient.

Jeff is also intrigued by projects that focus on creating enormous energy storage units, replacing batteries or other materials that can store heat to use it for industrial purposes or for warming people's homes on cold, wintry days.

Jeff's parents once heated their Lunenberg farmhouse by burning wood. They showed him how to split the logs. They also instilled in him the importance of study across disciplines. They taught Jeff the value of hard work, encouraged his great curiosity, and, when he had to repair the tractor or walk home, he learned resilience and honed his ability to figure out how to repair and redesign machines. Jeff says coming to Canada was the best thing that could have happened in his life, and Canada is indeed fortunate that he remains here innovating and inspiring his students.

43

ANNA KAZANTSEVA

Learning with Machines

Victoria, BC

The foundations for machine learning and machine translation are the work of computational linguists such as Anna Kazantseva, who uses her impressive skills and talents not to speak on behalf of people but to offer them the tools they need to preserve their language and culture. Working with communities that are in danger of losing their language through assimilation, she assists in creating written texts and structural and grammatical analyses, which she then writes in computer code to capture the language. Her work actually records living history and gives the spoken word the means to reach future generations. Anna's careful and painstaking transcriptions that she then encodes become useful for teaching and learning and provide access to cultural expressions for this and future generations. Anna is proud of her accomplishments, but says, "Anybody could have done it." She thoughtfully adds, "I am just one person." But to those who can now speak, read, and write in their native language, she is a brilliant, dedicated, and innovative researcher who provides a forum to preserve memories, and offers vital tools to support the creation of the stories the next generation may transcribe.

Born in Stavropol, a small town in southwestern Russia, Anna was encouraged by her family to study hard as she was growing up and to take advantage of every possibility to enrich her education, especially since there were no science courses offered in her school. When she was sixteen, they encouraged her to participate in an exchange program, which took her to San Antonio, Texas. She found her new environment quite a culture shock, being exposed for the first time to a place so different from her small town. She missed her family and sent handwritten letters to them, as they did not have access to a computer. They could only afford one telephone call each week. However, she became aware of the many doors that education might open and the experience made her want to learn more about the world. In 1999, she received a scholarship to study in France, which was at once a journey and a form of emancipation.

On entering the American University of Paris, she started a business degree and supplemented her scholarship with a student loan. During her second year, she asked to take a computer science course instead of the required Visual Basics for Business. This was the beginning of her new career. In four years, she completed two undergraduate degrees in business and computer science, and then followed her boyfriend to Canada in 2003. Due to the immigration rules of the day, she could not work, but was able to continue her education. This was for her a great pleasure because, for the first time, she did not have to combine employment and study. As soon as she arrived in Canada, she went to the University of Ottawa to do her master's in computer science and discovered the field of natural language processing. She went on to earn a master's in computing science and then completed her PhD in 2014 at the same university.

Upon graduation Anna went to the National Research Council (NRC) and began collaborating with a group on multilingual text processing. The group was working with many of the world's languages, including English, French, Arabic, Chinese, and Russian.

Anna had a new idea. Why not also work on Indigenous languages in collaboration with Indigenous people in Canada?

She began researching language-related technologies that might be useful in the context of Indigenous language revitalization and stabilization. Machine translation seemed like a "cool option," she says, but after talking to teachers and students in immersion schools and to experienced language activists, she quickly realized that her assumptions were unfounded. In most cases, teachers and students had little or no interest in machine translation.

But there was a need for tools to help students continue their learning at home; to aid teachers in speeding up the creation of curricular materials; and to help beginner students learn how to write and read more quickly. She soon realized that without the guidance of the Indigenous people, who for many years were already at the forefront of language revitalization, useful language technology could not be created. Dialogue was crucial. More than that, it was the act of listening and learning that was critical, as well as a deeply humbling and transformative experience.

In 2017, the NRC team, together with Canadian Heritage, was tasked with creating language technology that would help revitalize Indigenous languages in Canada. The project was a direct result of Anna's earlier, preliminary research.

Anna's own research has been focused on Kanien'kéha, which is the language of the Mohawk Nation. With the guidance and expertise of Kanienkehá:ka teachers, she has created a verb conjugator. Verbs are central to the language and very complex.

More recently, Anna has been part of a project that synthesizes speech for Indigenous languages to be used in educational settings. The pilot is for three very different languages: Mohawk, Cree, and SENĆOŦEN, the language of the First Nations Saanich people.

She notes that the field of language processing is fast moving and often in a direction that may not be the most needed by the population. For example, it is used for advertisements to sell products and to increase the speed at which emails can be answered. Anna says it can also be used to preserve cultural heritage. Language is the most significant means by which culture is transmitted, and ancestral languages require a great deal of respect. Our rushed culture of today makes it easy to hurry past words. It is important to consider their value and meaning. Words reflect ethical values and our times. And they are constantly changing.

Anna's work is fortunate to be grounded in decades of painstaking language study by Indigenous communities and individual scholars. For Anna, what is important is not simply accomplishing technical goals, but respecting the wishes, opinions, and needs of the users and their communities. Anna observes that her team is more productive, healthier, and better balanced because it includes Indigenous members. Over the course of this project, she has been fortunate to meet many inspiring people and has heard about many more. For instance, the late Dave Elliott, a towering figure in the SENĆOŦEN-speaking community, played a pivotal role in reversing the decline of the language. He single-handedly created a writing system that did not require expensive keyboards, made multiple recordings of elders and speakers, and produced a good deal of curricular material for the local language school. Much of this work was accomplished while he was working as a janitor. Owennatékha Brian Maracle, along with his partner, Audrey Maracle, founded the Onkwawénna Kentyóhkwa immersion school of Mohawk. The school went on to become world-renowned and highly successful in creating proficient speakers. Anna says that, previously, colonizers tried to relegate Indigenous peoples to the past and denied them access to modernity. She is working to provide this access and would like to see all Indigenous languages have the same digital support as languages like English and French.

Since Indigenous ownership of the software was the goal from the outset, to make this ownership meaningful the code had to be simple enough and sufficiently understandable not to require a highly specialized person to make changes. Anna's next task is to make her code even less complex so that people can just "take it and run," using it as they wish. For some tasks, it is feasible to make the code portable to other languages, which is highly desirable. She notes that there is a great deal of language revitalization around the world, and that while there is no single recipe for success, she is hopeful that the software created for this project will be a small step in the right direction.

Anna's advice to young people is to "really apply yourself. Having a good idea is important, but in and of itself it is not enough. You must be willing to do service and to have a genuine interest in what you do." Her own daughter is studying to be an artist, and Anna told her that she must work extra hard to succeed. And if she does, she will do well and will feel rewarded. You need to start by laying the foundation before you can build a building. The same is true for careers and dreams. Anna has given us hope to fuel our dreams and her innovative tools will serve to realize them.

44

LEYLA SOLEYMANI

A Shining Light in Material Science

Hamilton, ON

Leyla Soleymani is only beginning her career, but has been named to a Canada Research Chair in miniaturized and biomedical devices, reflecting her significant contributions in the fields of advanced materials, biosensing, and nanotechnology. She has also brought products to market, started companies, and is now associate vice president for research in commercialization and entrepreneurship at McMaster University in Hamilton, Ontario. Her boundless energy, combined with her unfailingly creative mind, are the keys to her success. When she sees a problem, she immediately seeks solutions and asks her favourite question, "What if?"

In the 1990s, Leyla's family moved to Canada from Iran, settling in Montréal. Although her parents encouraged her to explore different fields of study, she gravitated toward engineering. Inspired by her uncle who was a professor of engineering working on wireless communications, her father, who was an electrical engineer, and her mother, the computer scientist in the family, Leyla enrolled at McGill University in 2001 and became the fourth engineer in the immediate family, receiving a degree in electrical engineering. However, she was also inspired by physics and specifically the field of photonics, which is the study of the interaction between light and materials.

She particularly remembers Andrew G. Kirk, professor of electrical and computer engineering whose classes in engineering included the study of the history of science, thus situating engineering in a broader context. She was invited to do some research in Dr. Kirk's lab, where she discovered "cool toys," like photon detectors. She did what she now thinks of as simple experiments: counting and measuring photons and grading light that passed through various objects. The lab work convinced her that she wanted to go on to graduate studies, and a summer internship at Pratt & Whitney allowed her to understand firsthand the important links between research and industry, a relationship she would continue to explore throughout her career.

In 2005, Leyla attended the University of Southern California, where she studied electrical engineering and electro-physics, completing a master's degree in engineering. At that time, nanotech was an emerging field involving research into very small materials that could be developed and applied to industry. Leyla began thinking about nanotech applications that could transform everything from manufacturing to health care. She worked in four different labs and was involved in many projects, including making carbon nanotubes using a solution process. Nanotubes are largely cylindrical molecules created from hybridized carbon atoms and have a one-dimensional structure with enormous potential for application due to their combined flexibility, conductivity, and strength. They are used in flat-panel displays, bulletproof vests, aircraft, and transistors, for example. They could also potentially be used in targeted drug delivery and nerve-cell regeneration, an application that links to Leyla's ongoing interest in making health care more efficient and practical, helping both medical practitioners and patients.

On completing her master's degree in 2006, Leyla decided she wanted to do her PhD. She was accepted at several universities and chose the University of Toronto because of Ted Sargent, distinguished university professor in the Department of Electrical Engineering. Sargent had been a visiting professor at University of California, Los Angeles, where he gave a lecture that Leyla attended. He took the time to talk with Leyla afterward and even invited her to work in his lab, where he was doing research on solar cells. When he added a project using sensors in the field of health, Leyla agreed to work on it, thereafter dedicating her research to biosensors. She collaborated with Shana Kelley, a well-known professor of biochemistry, pharmaceutical sciences, chemistry, and biomedical engineering, who was working in the Faculty of Pharmacy. Leyla completed her dissertation on the electrochemical detection of DNA. This technology has an important application in rapid tests for infectious diseases. Traditionally, samples would be sent to a lab, where it would take twenty-four hours for the culture to grow and be identified. The new rapid tests provided immediate results, enabling physicians to prescribe the proper antibiotic during a patient's visit, starting treatment without delay. Leyla's work combines physics and engineering, and more specifically electrical engineering with electro-physics.

In 2011, Leyla was appointed at McMaster University, where she would be provided a lab to pursue her research. While the lab was being set up, she met with her colleagues and together they ended up doing some low-tech development work, looking at ways to automatically deliver drugs and insulin, ways to regenerate cells, and, finally, creating a process to coat surfaces with antibacterial film, using consumer materials that were readily available in the local hardware store. Normally, surfaces have to be lubricated to repel liquids. Leyla shrank films (think plastic wrap) to obtain nano-texturing, which are new structures characterized by nanoscale wrinkles that were highly repellent and also antibacterial. The new films could be spread out and fused to surfaces, which would then also be antibacterial, preventing the spread of infection. Surfaces in kitchens, serving areas, airplanes (armrests and tray tables, for example), trains, handrails on stairs and escalators, washrooms, and school desks could be built covered with this material. When the pandemic struck, Leyla and her colleagues tested the material again, this time specifically against viruses. The results were the same, and they published several papers describing their findings, which received significant media coverage as an innovative and practical way to limit the spread of contagious diseases.

Leyla says that innovation requires thinking outside the box. Throughout our lives, she says, we must be flexible. Living through challenging times in Iran, when she was confined to her home, she learned that one does not always have access to what one would need or want. It becomes necessary to use whatever materials are at hand and transform them, making the best of a situation whenever possible. When her lab at the university was not ready, she just used the materials available, looked at the raw data, and changed the established mindset about what is or could be possible. She constantly asks herself, "What if we did things differently?"

Today she is "really, really excited about wearable sensors and a continuous glucose monitor that would both provide the reading of the glucose level for diabetics and automatically inject the proper amount of insulin in a timely fashion." She imagines real-time sensors that can be implanted or even administered orally, which will be able to transmit information, and perform interventions, as well as detect a wide range of diseases. She is currently working on a cardiac patch that could be used by astronauts. She is passionate about her work, which will have a real-life impact.

Leyla's advice to students is to follow their passions and to be patient. Step back, ask, "What if?" and imagine how things could be created. Then use the scientific method and experiment. Start with interesting ideas, then put them together and test them. If they work, they may be life-changing.

45

PRITI WANJARA

A Super Innovator

Montréal, QC

Priti Wanjara is an innovator who knows how to solve problems, take brilliant ideas and get them adopted. She helps technology cross the infamous valley of death, that space that lies between the design of an idea and its adoption for production and actual use. She says, "I somewhat haphazardly found my own path and now I particularly enjoy bringing solutions to others."

Born in Mumbai in 1970, Priti moved, at the age of five, with her parents to Montréal, Québec. Her parents always supported her passions and interests, but it was her maternal grandfather in India who provided her the most encouragement. He had had six children, but could only afford to educate his two sons. Priti believes his interest in her progress, and that of her sister and cousin, was sparked by his regret at not being able to send all of his own children to university. He cared about her struggles, and through their handwritten correspondence he encouraged her to apply herself diligently and was delighted when she succeeded.

When Priti came to Canada, she could speak neither English nor French and struggled to catch up with her classmates. This changed in high school, where the students were introduced to the science and math curriculum. The language of science spoke to Priti and, suddenly, the playing field became equal. When it came time to decide on a topic to pursue in university,

Priti hesitated between health sciences and engineering. She decided to test herself by volunteering at the hospital and determined that medicine was not the perfect fit for her. She then worked at her CEGEP secondary school as a math and science tutor, which she greatly enjoyed. She also wondered if she should study physics, chemistry, or perhaps computer science and engineering? She believes that fate played a role in her decision. One day, she noticed a poster announcing an open house in the Faculty of Engineering at McGill with a lecture by the former professor Ralph Harris, an expert in extractive and process metallurgy. She decided to attend. After Professor Harris gave an excellent presentation on materials engineering, Priti says she "was sold." In 1989, she went to McGill and completed both undergraduate and graduate degrees in the same field.

After graduating in 1999, she worked at Ivaco Rolling Mills in L'Orignal, Ontario, where she developed innovative steels and their processing to hot-rolled end products. After the birth of her first son in 2001, she was laid off. But as one door closed, another opened: she interviewed for a research position at the National Research Council of Canada (NRC) and was hired to work on the forging characteristics of titanium materials.

During her early days at NRC, when she was trying to join strips of titanium together, she thought about how melting and welding are applied in the field of aerospace. She expanded this research to metal 3D printing, also known as additive manufacturing, which is an innovative process for layer-by-layer manufacturing to build objects. Since then, she has used her metallurgy skills to solve a wide range of materials and process-related challenges. Today her work at NRC involves studying the material-process-structure-performance relationships of 3D-printed metals to support their development for a wide range of applications in the aerospace, defence, automotive, and power generation sectors. She has also deliberated on the possibility of metal 3D printing in space using in situ resources from the moon, for instance, to print lunar-derived alloys and create physical infrastructure on the moon that may be built and serviced robotically with minor human involvement. Currently, she is collaborating with researchers at Carleton University to demonstrate a complete processing chain that will start with raw, lunar mineral analogues and end with usable products that could actually be manufactured one day with 3D printers by people living on the moon. While some scientists work on travel to the moon, Priti and colleagues are thinking even further ahead. They are contemplating the day when people might actually be living on the moon. They will have to be able to manufacture needed machines on-site with the lunar materials available.

Priti also works with small- and medium-sized businesses (SMEs) to develop appropriate solutions. Too often SMEs grow to a certain size, but then cannot continue due to technical problems. Their early progress toward success simply stops and they become mired in the sands, aptly named "the valley of death." She offers one example of the many problems she has solved: Cars need to be protected from corrosion and this is habitually done with a zinc coating. To extract the zinc for these coatings, an electrowinning process is typically employed using nearly pure aluminum cathodes that deteriorate with time and are expensive to replace. Priti saw an opportunity to repair the cathodes by friction stir welding (FSW), which is an advanced, solid-state joining process that uses a nonconsumable tool to join materials without melting. She then led the development from small lab-scale prototypes to full-scale cathode repair for demonstration at Canadian Electrolytic Zinc (CEZinc), the largest refiner of zinc in eastern North America, which had been established in 1963 as a subsidiary of Glencore in Québec. The FSW solution has proven to be both robust and profitable. Since implementation in 2015, roughly ten thousand cathodes are repaired annually by FSW at Groupe Tremblay, and the use of the automated elements has brought considerable operational savings by cutting their production time by 80 percent. It has also contributed to a greener environment with the reduction of primary metal waste by 35 percent and the scrap rate by 50 percent.

Priti is now working on two joint international projects between Canada and Germany involving the implementation of artificial intelligence to streamline process monitoring and quality-control measures for smart/digital manufacturing of products. One project aims to develop advanced AI-based software to automatically run Directed Energy Deposition (DED) 3D printers. A second project aims to develop an AI-based quality-control system for automotive production plants, providing real-time analysis of the joint integrity enabling modifications to be made on the spot, thereby reducing the scrapping of parts.

Today Priti is an adjunct professor at McGill, offering unpaid lectures, inviting students to visit her lab, and extending the invitation to students at Concordia University in Montréal and the Royal Military College in Kingston. She wants to "pay it forward," by offering some inspiration to the next generation of researchers, just as Professor Harris did for her. Priti offers the following advice for success: hard work and perseverance are important, but you really need to discover your passion first.

"Pursuing a STEM career is challenging and requires dedication and sacrifices," Priti says. "You need to do research and face different challenges each day, keep up with the literature, publish articles, push boundaries, and think creatively. But, if you are passionate about science, you will embrace the lifelong learning with eagerness and continue exploring after each failed experiment until you succeed. If you do not love what you do, your motivation will wane over time."

Priti's views on pursuits for happiness come from her late mother, who did not have the advantage of an education and was a homemaker and seamstress. Nonetheless, her mother was a powerful, insightful person who inspired and directed Priti to pursue a career path that would make her happy. Young researchers often ask her how she manages to balance her career and family. Priti says, "I am no 'Superwoman.'" She does not believe that Superwoman exists. Priti says she once thought she should be like her mother, who was self-sacrificing and did all she could to support her family. She soon realized that it was impossible, and she attributes her ability to succeed at home and at work to her biggest fans: her sister, father, in-laws, and her husband.

"Everyone needs a community, a network of support," Priti stresses. "We each have our life to live and our story to tell." Priti's story includes many successful innovations and more to come.

46

TEODOR VERES

Inventing the Future

Montréal, QC

What Teodor Veres likes best in life is to invent technologies. It is not about the number of patents (and he has over two hundred); it is about helping people, saving lives, and supporting colleagues.

"There is no bigger pleasure than ideas firing in my mind and getting enhanced by colleagues," he says.

In 1962, Teodor began life in Malin, a small village of 4,200 in what was then Transylvania, and is now Romania. His family farmed and did not have formal schooling. But they understood that education was the only way to have a better life, and they gave Teodor the opportunity to learn. Indeed, the population in general respected educated people, and to be a professor in a university was deemed the pinnacle of success.

In 1969, Teodor finished elementary school, and, as there was no high school in his hometown, he had to leave home and go to Nuseni to attend high school. There, he met a math teacher who took Teodor under his wing. This teacher opened up his library to young Teodor, who remembers it as "a wonderland with thousands of books." On seeing this wealth of knowledge, Teodor realized that "everything is possible."

Perhaps inspired by the space launch, which also captured his imagination and made him dream of being a scientist who might contribute to space travel, Teodor applied himself to his studies, excelling in math and physics, and qualifying for the national Science Olympics team, which made it to the finals. This was a highly motivating experience that gave Teodor the ambition to attend university. At that time, young men either went to university or had to serve in the army for two years, after which they might no longer be able to attend university. The university exam was the turning point in life: Teodor passed and went to Cluj-Napoca University on the northern border of Romania. There he completed his studies for a degree that was the equivalent of a master's during the last two years of the Communist regime. As part of the scholarship he had received, Teodor had to teach physics for three years at a very small agro-industrial college. He thought that this was the end of his career. However, he did marry and start a family during that time.

In an attempt to break out of this intellectual isolation, he met with some of his former professors at the university, who invited him to a competition in 1989. As a result of his extremely impressive scores, he was named assistant professor of physics and was awarded, in 1992, a scholarship to do his PhD in a joint Romania/France program. He conducted the research part of his program at the Louis Néel laboratory in Grenoble, part of the Centre national de la recherche scientifique (CNRS) in France. Once again there was a price to be paid. France would not allow him to bring his family. He had to leave his wife and his two-year-old son behind. When he returned to Romania in 1994, things were changing with the influx of Russians and the fact that there were few resources for higher education and even less for experimental physics. It was clear that the family had to leave. They decided to immigrate to Canada. They landed in Montréal in 1994 and were interviewed by an immigration officer, who told them that Teodor's wife, who was a computer scientist, would have no trouble finding a job. Teodor, on the other hand, with his degrees in physics, should be prepared to become a dishwasher.

Indeed, Teodor could not find work, so he applied to the Université de Montréal to do another PhD in physics. He was awarded a scholarship of $13,000, which was just enough to feed his family. When he completed his PhD, the Natural Sciences and Engineering Research Council (NSERC) offered him a postdoc, but it had to be used to study abroad. Teodor did not want to leave his family again, so he started looking for employment. One day, his wife noticed a building with the Canadian flag on it. Teodor learned the building housed labs run by the National Research Council (NRC) and that they were looking for a researcher in nanotechnology. He applied and was offered the job, which was not really in the field in which he wanted to work, but he accepted anyway. He was also not really happy, as he did not see a career trajectory for himself. He was in the process of applying to teach in a CEGEP when the director general of the NRC visited the nanotech lab, met Teodor, and said that he could work for two years on the projects of his choice, but he would have to come up with a way to finance his research on his own. When Teodor heard "Do whatever you like," he could not believe his ears. He had been "given the chance to jump out of the plane without a parachute, but high enough to be able to choose a landing place."

He found the financing and immediately started looking at building new methods of creating industrial polymers. He likens this technology to creating the Gutenberg press with a replication feature for sensing cells. While the Gutenberg press copied and made written texts widely available, the new methods of creating industrial polymers would make a previously rare material easily obtainable. Teodor's specific topic was using nanoscale printing for photography. He worked with polymers to develop tiny objects that would automate the chemical analyses done in biomedical labs in order to reduce the cost and increase the speed. The objective was to make molecular diagnostics accessible through the miniaturization of computers and medical instruments. What he created is a true innovation: a "lab on chips."

Teodor now has forty-two people working with him. He is well published and highly regarded. He has contributed to the Canadian Space Agency's work and realized his childhood dream, inspired by the moon landing, of contributing to the science behind space travel. The instrument he developed is now being tested for use at the space station.

Teodor says the most successful products are not the biggest scientific advances, but ones that affect people's lives. He thinks we must go beyond our competitive instincts as scientists and collaborate. He himself has made "unholy but successful" alliances with his biggest competitors. He says, "70 percent of biotech engineers are leaving the country, and we import

70 percent of the medical devices we use. In Canada the scientific community is small. Why not collaborate?" He thinks it is possible and not too late to make Canada internationally competitive in science and technology.

Today he is cogitating on how to use molecular science in genomics. What should be done with all the data produced? He is working on an invention for hospital intensive care units, which will triage patients by blood analyses, determining automatically who needs to be placed on a ventilator.

His advice to young people is the same that he gave his son when he won the US Science Fair: to use his talents and education to drive innovation and to help others. He encourages schools to give students the opportunity to get hands-on experiences in real, working labs. He himself has personally rented a car to bring instruments to schools for demonstrations, and he has a nanoscience program through which he has tutored two thousand students.

Teodor sees the future as being strongly influenced by quantum processing and artificial intelligence. The way this capacity is applied in biological and medical science will be increasingly important. There will be a shift in the way we design and prescribe medication using information technology. In the future we may be able to decentralize manufacturing and produce and deliver the medication close to the patient at the specific moment that is appropriate to the pathology.

Teodor says the pandemic has taught us that we need to use science to take care of people. He maintains that hope and caution for the future must go hand in hand, but we know that, with Teodor at work, we have great reason for hope.

47

ANDREW PELLING

Thinking out of the Box

Ottawa, ON

Fellow of the Royal Society of Biology, fellow of the Royal Society of Canada, TED Senior Fellow, founder of many start-up companies, and holder of a Canada Research Chair, Andrew Pelling has invented a new, disruptive pathway to innovation.

His story starts in 1978, when he was born just outside of Hitchin, a town on the outskirts of London, England, and continues four years later when his family moved to a city just outside of Toronto, Ontario. His dad, an engineer, met his mom, a computer scientist, by coming gallantly to her rescue when they were students. The only woman in her class, she was often the object of pranks. When her fellow students threw a water balloon at her, Andrew's father provided her with towels and encouragement, which soon led to a romance, marriage, and young Andrew.

He cannot recall a teacher in high school who truly inspired him, but he does remember the chemistry teacher who told him that he should not contemplate a career as a chemist, after he failed ninth-grade chemistry. His parents, however, supported a broad education and encouraged him to pursue the study of science. Perhaps this encouragement combined with his teacher's comment spurred his desire to excel in chemistry and to earn a PhD in the field.

Andrew spent the summers during high school working in a salad factory, packing boxes and delivering salads. That job motivated him to attend university. From 1997 to 2001, he attended the University of Toronto, where an older friend advised him to enroll in a chemistry course for overachieving students. On the first day of class, the professor won Andrew's undivided attention by announcing, "At the heart of any good chemist is the desire to make things and blow things up."

Andrew was passionate and curious. He asked a lot of questions and liked using the scientific method and tools in the lab to find answers. He admits that he probably spent more time in the lab than in his classes, and even as an undergraduate he was publishing scientific papers. He was encouraged by John Pezacki, a passionate and dedicated postdoctoral student, whom he would meet again when they were both professors at the University of Ottawa. Andrew says he learned from John how to be a scientist. He also says that John paid for so many coffees back then that he tries to treat John whenever they meet now.

In 2001, Andrew elected to do his graduate work at the University of California, Los Angeles, where he was delighted by the interdisciplinary culture that enabled collaboration with the visual arts. He recalls having had the opportunity to organize a large, ten-thousand-square-foot nanotechnology and art exhibit. When he visited an artist's studio, it reminded him of his own lab with space for experimenting and space for designing applications of ideas. He realized that there wasn't such a great difference between the arts and sciences. Both are rooted in curiosity.

For his thesis, Andrew demonstrated that a greater meaning and understanding of the whole of any organism can be obtained by describing the organism in a way that goes beyond any single sense of perception. For example, if you could both see and feel a cell, you could discover its state of health and well-being by its movement and its tactile qualities, such as hardness or softness. He created a device similar to one that bioengineers use to measure sound waves that could be correlated to physiological states using images on a screen and volume and frequency changes being both heard and graphed. He also posed some fundamental philosophical questions. At the time he was writing the thesis, the field of mechanobiology was in its infancy, and biochemical engineering had not yet become popular. Only a small number of researchers worked in these interdisciplinary fields.

From 2004 to 2008, he held a postdoc at University College in London, where he had a wonderful supervisor who not only loved novel ideas, but also let Andrew start an independent career. His first job was at the London Centre for Nanotechnology at University College, but he realized that he would best flourish in the freedom offered at a smaller university. He wanted to find a place where knowledge was valued for itself, and not simply as a commodity that could be sold. In 2008, he was offered a Canada Research Chair at the University of Ottawa and was subsequently named a Tier 2 Canada Research Chair.

On arrival in Ottawa, he decided to create a communal space for students coming from every field of study and having a great diversity of experiences and perspectives. He wanted students to ask questions that had never been asked before, and he wanted them to use the scientific method to find answers. He began by training his students how to ask questions, identify solutions, and make new discoveries. While the goal of Andrew's innovative pedagogical methodology (which may well have had roots in Greek teachers like Socrates and Hippocrates, who dialogued with students asking questions) was the search for knowledge, the result was that he and his students filed more patents, published more papers, received more awards, and created more companies than some entire departments that concentrated their energies on the development and applications of technology.

The secret to Andrew's success is, he says, to "Let it happen." Do not set out with the conclusion; begin with a question. Do not constrict yourself to winning. The assumption of failure is freeing. Do as much as you can in the time frame available. His mantra is "Ideas are a dime a dozen." Verbalizing them is a bit harder, as is employing the scientific method to determine the validity of the idea.

"There are many dead projects out there, all along the way," he says.

Andrew demonstrates to his students how and why it is so important for curiosity to be practiced. He will build a strange device or pull another apart. He is an aficionado of dollar stores and of bad science fiction musicals like *Little Shop of Horrors*, which inspired him to build scientific equipment out of garbage. He built, for example, a cell culture incubator from discarded items, and his design has since been replicated thousands of times around the world, proving that science "can happen in areas without sufficient resources." The 3D printer he and his students created is currently operating, and the multicolored pink and orange microscope they built has been prominently placed in Canada's Science and Technology Museum as an example of innovation.

During the pandemic when students complained that they could not operate their microscopes from home, Andrew found a way to hook them up to Twitter and created software to make it possible to control the microscopes' functions remotely. He hypothesizes that one could possibly use social media platforms to build virtual labs, a bit like in the Minecraft game.

Now, let us return to *Little Shop of Horrors*. In the 1986 film adaptation, actor Rick Moranis owns a flower shop where a carnivorous plant is growing out of control and eating people. The plant is part mammal and part vegetable. This idea prompted Andrew to wonder if human cells could be propagated on a plant leaf. The experiment failed due to the waxy coating on the back of the leaf. However, one of his students thought about peeling an apple to remove the waxy skin and wondered if that might allow sufficient moisture to foster cell growth. Andrew peeled the skin and removed the identifying DNA from the apple cells and injected the DNA of mammalian cells from a human ear. The experiment worked, and they published their findings to what Andrew describes as "much criticism and ridicule," with the exception of a company in Australia, which is now working to bring the concept to market. After all, they had broken every rule of tissue engineering. Andrew has since experimented with other plants.

Andrew's next innovation is inspired by the film *The Matrix*. In the film, the body becomes a battery for systems. He is thinking about the interface between living tissues and electronics. Can we draw power from tissue constructs? If we grew tissues in the lab, and they created their own power circuits that would fuel them and keep them alive, it would be a symbiotic relationship.

Andrew remembers an offhand comment he made about creating space for innovation. He says that today's post-pandemic world of work is less about creating one single space and more about finding the method to spark creativity and to engage resources that are distributed among partners at a distance. One needs to reengage and curate teams in different places. The framework should include milestones, measurable deliverables. An agile team that has the skills and techniques to measure failure as well as success is essential. His idea to focus on the method for sparking creativity, as opposed to the environment, took off like "wildfire" and has been successfully applied in many creative corporate environments.

For example, Andrew holds competitions requesting submissions for the most outrageous projects and will select two or three applications out of seventy. An idea bound to succeed is considered too safe and is put aside. The winning ideas are those that are located right on the borderline between possibility and failure—this is where innovation lies and exactly where Andrew likes to work.

PART EIGHT

Physics, Photonics, and Nobel Laureates

Physics is a field of science that is generally about problem-solving, beginning with basic existential questions such as: What is the origin of the Earth and the universe? What are the tiny particles we cannot see? If light is energy, can we harness it for our use? What are the forces, like magnets or gravity, that attract or propel things forward or prevent them from moving?

Researchers in the field of physics design experiments to determine the nature of motion, energy, and force to be able to harness them to improve our lives. There are many branches of physics that combine with other fields such as chemistry or engineering, medicine or astronomy. Astrophysicists, to take one example illustrated in this chapter, hypothesized that there were minuscule particles, even smaller than electrons, in our atmosphere. They called them neutrinos and imagined that 100 billion of these invisible particles could pass through one's thumbnail every day without our knowing or feeling anything. Then physicists had to prove that neutrinos actually existed. They designed a series of experiments that would have to take place far underground where the air would be filtered by several kilometres of earth. In early days they posited that if they could pass a laser beam through purified air and heavy water, it would strike a neutrino. This actually occurred, and the scientists were thus able to confirm the mass of neutrinos by measuring the amount of energy they emitted on being struck. In later experiments for which Art McDonald won the Nobel Prize, they determined that neutrinos were not uniform and described them as being of three sorts, or "flavours." The neutrinos emanate from the sun and researchers counted the number of neutrinos that would reach the Earth. This, however, did not seem to be the case when they tried to measure them. The numbers simply did not seem to add up! Art and his colleagues demonstrated that the number calculated was indeed correct by discovering that there were different kinds, or "flavours," of neutrinos and showing that the overall number of neutrinos was the same, while the combination of flavours differed. Their conclusion is pure, basic science, but the way they arrived there was completely innovative and included setting up their lab in a working nickel mine far beneath the Earth's surface. As knowledge is gained from this work, researchers will discover ways to apply it to new ways to improve and support our lives.

Other astrophysicists hypothesized that there were other planetary systems in the universe and that our solar system was perhaps one of many. Proving that there were planets beyond our solar system (exoplanets) became another problem to be solved. Christian Marois armed himself with the most powerful telescopes possible and invented new ways to photograph planets that lay beyond our ability to see. He hunted the night skies and ended up with the first photograph of an exoplanet, proof that there are indeed other planets in space that might contain or enable life. He opened the door to the next questions that may one day offer humankind the possibility of locating additional means to support life, and to discover, and perhaps join, distant neighbours.

Optics, or the science of light, is another field of physics and photonics, a subfield of optics, can be described as the science of making and manipulating light, using it much as one might employ electricity. The advantage of light is that it uses less power and energy. Its applications are numerous and include fiber-optic networks for communication, the cameras in cell phones, bar codes, printers, DVD players, medical devices, agriculture, sensing, and imaging. These applications of light have changed our lives in a few short years. They required serious thought and concentrated effort to create powerful lasers and to control the speed and focus of light beams. Paul Corkum first imaged the theory and then created the lasers and the extremely rapid (attosecond) pulses. Donna Strickland's research, for which she received the Nobel Prize, led her to find a way to sequence pulses, allowing them to expand in intervals that would result in extremely precise lasers for use in optical surgery, for example. Stephen Mihailov was inspired by Paul's laser machine, which allowed him to pursue this research further and imagine new applications, including environmental sensing, the gasification of coal to produce hydrogen and carbon dioxide, which can then be sequestered, the electrification of vehicles, and improvements to jet engines.

Researchers in physics are innovative in positing theories that might explain the macro and micro problems we do not understand. They then design experiments that could prove or disprove the theories. Finally, they are innovative in finding ways to use the new information about energy, motion, and force to create tools that will improve our lives and enhance our ability to solve further mysteries.

48

ART MCDONALD

Astrophysicist, Innovator, and Nobel Laureate

Sydney, NS; Kingston, ON

Sydney, Nova Scotia, is home today, as it was in Art McDonald's youth, to around thirty thousand Islanders. The steel mill has closed, the tar ponds have been cleaned, and the town once known as the industrial hub of Nova Scotia now relies more on small businesses and tourism. Yet, over the years, the same close-knit community continues to thrive.

Art's story began perhaps with the early Scottish and French settlers who came to Canada in the late eighteenth and early nineteenth centuries, and who had to reinvent themselves and their way of life in a new land. More than half a century later, Art's family history continued with his dad, a lieutenant in the Canadian Army who was wounded in the Second World War. Returning home in 1946 with the Military Cross to his wife and young son, he became a city councillor and Art's mother volunteered in the community. Both encouraged Art in his studies, as did his teachers. Art recalls Bob Chafe, the math teacher who gave extra classes beyond the curriculum. Like many young scholars, Art had a paper route that he recalls with some humour as being "uphill all the way," as he balanced a heavy load of papers on his bicycle. He attended the HI-Y, a service club for teens that raised funds for charity and held youth dances after Friday night meetings. Rock 'n' roll was all the rage and Art loved to dance. That is how he met Janet, whom he would later marry.

In 1960, at the age of seventeen, Art left Sydney for Halifax, Nova Scotia, to study science at Dalhousie University. There, he learned that mathematics "could be used to figure out how the world works." He liked solving physics problems and got a summer job through his professor measuring gravity on roads in Nova Scotia. The research team discovered an anomaly that pointed to the presence of gypsum, a mineral used for fertilizer, chalk, drywall, and other materials, which in turn led to the establishment of a mine. Art and his friend the future economist, politician, and science policy expert Peter Nicholson later persuaded their professors to fund a project to create a file of information for students who might be interested in going to graduate school and wanted to know in those days, before everyone could simply search the web, which fields were available at each university as well as the cost and opportunities for scholarships. In the process, they learned about and applied to the California Institute of Technology (Caltech) and Stanford, respectively. Both were accepted and had extraordinary opportunities to study.

On arriving at Caltech in 1965, Art made his first major discovery in the basement of the Kellogg Laboratory: the Van de Graaff accelerator, which accelerates subatomic particles to great speeds and is used mainly in research. Under the direction of Charlie Barnes, who was his thesis supervisor and mentor, an expert on the nuclear reactions that produce many of the elements of the universe, and one of the founders of the field of nuclear astrophysics, Art was able to learn about the basic laws of physics through firsthand experiments. He was also inspired by American nuclear physicist and astrophysicist Willie Fowler, who later won a Nobel Prize for explaining the nuclear reactions that power the sun. Two of Art's fellow students went on to continue working in astrophysics, and one, Hay-Boon Mak, became a professor at Queen's University and would later conduct experiments at the Sudbury Neutrino Observatory (SNO). Art ended up eventually following Mak's path. In 1969, Art graduated from Caltech with a PhD in physics and was immediately welcomed at the Atomic Energy of Canada's Chalk River Nuclear Laboratories, where he continued research on topics that included isotopes and the forbidden decays in light nuclei, the forces that hold nuclei together and the energy released at their decomposition. One of the highlights of this period involved using a high-intensity electron beam to measure the disintegration of deuterium exposed to gamma rays. This experiment was possible due to an innovation. With his colleague Davis Earle, an expert in high-energy physics, Art created a new, continuous-beam polarized electron source. After their work concluded, the source was transferred to the electron accelerator at MIT for use in further experiments in studying the energy and particles released by the collision of high-speed electrons. From 1982 to 1989, Art worked at Princeton University with a team on polarized targets, which involved scattering experiments to study the nucleon spin structure of protons, neutrons, or deuterons (the nucleus of deuterium). Their approach included new techniques enabling nuclear and particle physics measurements. These innovations have since been applied to medical imaging.

The passion for precise measurement next led Art to Sudbury, Ontario, where a monumental project would be conceived and realized. In 1984, Herb Chen, a theoretical and experimental physicist who worked on neutrinos, and George Ewan, an expert in nuclear physics who was also studying solar neutrinos, began contemplating how to eliminate atmospheric contaminants such as cosmic rays from their laboratories and fixed on building a lab in the hollowed-out cavities of an operating nickel mine where, having largely eliminated contaminants, they could obtain the most pure and correct measurements of specific particles. Today when science and business collaborate, they are celebrated; but at the time they were not immediately hailed as extraordinary innovators. Herb and George began the work with Art and a dozen others. And in 1985, they were joined by David Sinclair, a world-renowned particle physicist. Together, they built the Sudbury Neutrino Observatory (SNO), the world's largest, cleanest laboratory that, at two kilometres underground, was also the deepest laboratory in the world. It took five years of planning and fundraising before they could begin construction. By that time, more than seventy scientists from fourteen institutions in Canada, the US, and the UK had become involved in the project, and in 1991 Art McDonald became the director of the SNO Institute, responsible for the lab, and SNO Collaboration, the network of scientists participating in experiments at SNOLAB. They also managed to borrow one thousand tonnes of heavy water to be used in their experiments and built a sphere large enough to contain it. The sphere was composed of 120 parts that had to be transported in the mine elevator and reassembled in the underground lab, a sterile environment that required all the floors and walls of the mine cavity to be painted and the air filtered to ten thousand times cleaner than the mine air.

The elevator, or "mine cage," makes the two-kilometre descent early in the morning and the number of passengers is limited. Scientists stand shoulder to shoulder with a few dozen miners going down for their shifts. All don heavy yellow overalls, boots, hard hats, headlamps, and safety goggles. The elevator's speed goes up to forty kilometres per hour, but for those unaccustomed to traveling 2,070 metres underground, it seems like an eternity. When the elevator doors open, the passengers step out into a dark tunnel carved out of rock and walk 1.5 kilometres to the lab, where around forty scientists at any one time work efficiently in a five-thousand-square-foot warren.

But what was it exactly that inspired researchers to dedicate years of effort, governments to spend over $70 million, miners to share their space, Atomic Energy of Canada to lend the project those one thousand tonnes of heavy water (which, by the way, have been returned), and hundreds of scientists to spend their time so cheerfully underground?

Human beings have always wondered why the universe exists. The answer always seems to lie somewhere between the infinitely distant (light-years away) and the infinitesimally small.

In 1930, Austrian physicist Wolfgang Pauli postulated that there were subatomic particles, but he thought they were without mass and would never be identified. Since then, these particles were proven to have mass and have been named "neutrinos," then thought to be the tiniest particles in the universe, coming from the sun and supernovae. The sun is powered by nuclear reactions that give off billions of neutrinos every second. They surround us, pass through our bodies, and even through the two kilometres of rock that buries SNOLAB. Physicists created models predicting the number of neutrinos that would be produced by the sun and arrive on Earth. Scientists at SNOLAB first set out to capture neutrinos in a detector where they would be stopped or scattered by the heavy water contained in the sphere. The neutrinos produced small flashes of light, and because they were passing through heavy water, scientists could identify different types of neutrinos. Three distinct flavours have been detected: electron, muon, and tau. Art and his team demonstrated that the electron neutrinos underwent oscillations and became muon and tau, a process that implies that they have a finite mass. These observations required changes to the Standard Model for Particle Physics at the most fundamental level and were the basis for being awarded the Nobel Prize in Physics. The SNOLAB results also demonstrated that the number of neutrinos—electron, muon, and tau together—equaled the total number predicted by the earlier theoretical models. Art and his colleagues proved the theory correct. This method could be extrapolated and used to inform the theories that surround nuclear fusion, which could provide one solution to the world's need for energy.

Today, Art and his colleagues continue to imagine experiments that will elucidate the enigmas of the universe. They have identified an ideal element to study rare radioactive decay: the semimetallic tellurium, which is found more commonly in the universe than on Earth. Fortunately, it is also found near Sudbury. One of the current SNO experiments is now called SNO+. The race continues to determine the absolute mass of neutrinos, and whether they are their own antiparticles (assuming that all particles are balanced by antiparticles).

Each of these experiments involved measurement, required years of patient effort and significant investment to set up, and led to discovery of information about life and the universe. And each necessitated a good deal of innovation: finding the right materials and constructing labs in unusual locations in difficult conditions.

Art and his colleagues at SNOLAB did not stop innovating when the COVID-19 pandemic put human lives at great risk. When the media reported that ventilators were in short supply around the world, they set aside their experiments and collaborated virtually with researchers at six Canadian institutions and agencies, as well as institutions and agencies from Italy, Germany, France, the US, and Switzerland. Together, with particle physicist Cristiano Galbiati at Princeton University and colleagues working at the Global Argon Dark Matter Collaboration in Italy, they developed a ventilator that could be built with inexpensive parts readily available off the shelf from sources around the world. They did not file for a patent, but offered the machine as a gift to humanity in a time of crisis.

Every time one converses with Art or hears him speak, he is quick to recognize others. On returning from the Nobel Prize ceremonies in Stockholm, Sweden, where he had negotiated to bring as many colleagues and their spouses as possible, he headed for Carleton University to thank David Sinclair, who spent his career at SNO and continues to lead experiments there and at the university, along with his colleagues and students. Every scientist in the building was crowded into the atrium, and there they stood, applauding and applauding and still applauding more, with genuine pride and pleasure. But Art's mission was to thank *them* for *their* part in the project. These students were involved in building parts for use in the experiments at SNOLAB, connecting virtually to labs across the country and around the world. Thereafter, Art travelled north to SNOLAB to congratulate the director, Nigel Smith, and assistant director, Clarence Virtue, and their colleagues both on the surface and underground. And before he made his way across the country, he returned to Queen's University, where he did not forget Mark Boulay, a professor of physics at Queen's (now at Carleton), and the students whom he himself had taught.

Art is busy paving the way for the next generation of Canadian innovators. As he collaborates with top researchers around the world, he opens doors for his colleagues, co-authors, and those students gathered in awe at his accomplishments. Art emphasizes that he never works alone. His team, his community, and his country support him and, in turn, all are inspired by his work.

49

DONNA STRICKLAND

Laser-Focused Nobel Laureate

Waterloo, ON

Thirty-three years after having written an article titled "Compression of Amplified Chirped Optical Pulses," Donna Strickland received an early-morning call from Sweden that changed her life: the Royal Swedish Academy of Sciences was recognizing her work with the Nobel Prize. Coming unexpectedly, the award interrupted Donna's career as a researcher and teacher and placed her on the world stage, where she found herself admired not only for her brilliant scientific work, but also as a role model for women considering careers in science.

Donna's life story began in the Maritimes. Her grandfather, a fisherman, was born in Newfoundland and had no formal education and learned to read and do math only after he retired. The Strickland family moved just across the Strait of Canso to Port Hawkesbury, Nova Scotia, where the local high school was so small it did not offer the final year. Donna's uncle managed to go to university despite the fact that he could not complete the final year of high school in Port Hawkesbury. However, two years later, when Donna's father attended the same high school, the final year had been added and his admission to university was easier.

Donna's mother, Edith Ranney, grew up in a farming village in Ontario and got a degree in teaching. Donna's father was an electrical engineer and moved to Guelph to work with the General Electric Company of Canada. He met Edith, a teacher, on a blind date. They married and had three children, the second being Donna.

Donna was born in 1959 and raised in Guelph, Ontario, but the family spent holidays in Cape Breton, visiting family and her father's home. Her parents "encouraged" her to go to university.

"It was not *if* you will go, but *when* you will go," she says.

She did not need the encouragement; ever since she was in elementary school, she knew she wanted to earn a PhD. She loved school and was always happy to start a new grade after summer holidays. Surrounded by her friends and siblings, she enjoyed her studies, and in particular math and science. She went to the local high school, where her mother taught English and history. She generally did very well, but she remembers clearly the one time she submitted a terrible job on a science fair project.

"You are so much better than that!" her grade thirteen teacher admonished her. She recalls several wonderful teachers, including her chemistry teacher, who gave the students Mars bars. She liked music, played clarinet, and was a member of the school band. She enjoyed camping and joined the Outers Club, which took the students camping in both winter and summer. When she won the grade eleven physics prize, she was concerned that her reputation as a "nerd" was sealed, but instead, she was warmly applauded by her peers.

During her school years she delivered newspapers, did some babysitting, and worked at the age of sixteen in a Fotomat booth, where film could be dropped off for processing. On a visit to the Ontario Science Centre, when she was still in elementary school, her father pointed out a laser to her, resulting in the family lore that he had inadvertently introduced her to her future career.

After graduating from high school, she decided not to follow her sister and friends to the University of Waterloo, wanting to strike out on her own and overcome her shyness. She could not decide between engineering and physics, but on finding that McMaster University, in Hamilton, Ontario, had a program that combined the two, she immediately applied and was accepted. When she arrived, she discovered that a quarter of the program was dedicated to lasers and electro-optics, the field that was to become her passion. In her second and her third year, she got a summer job in the laser group lab at the National Research Council (NRC) lab investigating and polishing fiber bundles, which involves sharpening the ends to optimize their performance in transmitting light.

She graduated in 1981 and wanted to continue studying lasers and went directly to the University of Rochester in New York State, where she earned her PhD in optics in 1989. Shortly after she arrived, a fellow Canadian student gave her a tour of campus and pointed out Gérard Mourou's lab. She subsequently met Mourou, a pioneer in electrical engineering and lasers, and began working in his lab, becoming one of his first PhD students when he was promoted from the research group to professor. Together, they succeeded in creating ultra-short laser pulses without destroying the amplifying material by stretching the pulses in time and reducing their power. Then they amplified the pulses and ended up compressing them again, which caused the intensity of the pulse to increase significantly. This is called chirped pulse amplification, and this was the discovery for which they shared the Nobel Prize more than three decades later. The very short and intense pulses of light beams can be used to make extremely precise incisions and are used in optical surgery and micromachining.

While working in the lab she met her husband, Doug Dykaar, who was a graduate student in electrical engineering. They continued their research programs and married some five years later in 1991, after Donna completed her postdoc and work at the NRC. She had wanted to work with Paul Corkum, who is internationally known for developing a new field of attosecond science, and she was a research associate in Corkum's lab in the Ultrafast Phenomena Group at the NRC from 1988 to 1991.

Donna then worked at the Lawrence Livermore lab in California in 1991 and subsequently at the Princeton Center for Photonics and Optoelectronic Materials in 1992, before returning to Canada in 1997 and a position in the Department of Physics and Astronomy at Waterloo, where she continues to inspire students and carry on a first-class research program.

Donna's research on lasers was delayed by the responsibilities of a Nobel laureate to speak on behalf of science and then by the COVID-19 pandemic. She has a number of projects on which she is working and is particularly interested in medical applications of lasers that would improve on the linear accelerators (LINACs) used in hospitals to direct radiation at tumors to shrink them.

Donna says her international experiences have changed her outlook on how science should be done and supported. She notes that South Korea has built a truly impressive "science city" in Daejeon, which is a technology and research hub bringing industry and academics together. In the Samsung building, she saw holographic televisions and other devices that were not destined for the mass market and noted that industry had been persuaded to work with academics on curiosity-driven exploration rather than on applied or mission-driven research, which is most often the case in North America.

Donna concludes that we need to find companies to invest in scientific research on campus and to increase grants for students who will be the workforce that will attract industry. Lasers provide a most promising future for environmental projects, and for work requiring sensing and infrared applications.

For more than three decades, Donna has been on a path of discovery and innovation. She modestly calls herself a "laser jock." But the truth is that she innovates in developing the equipment, conditions, and experiments that lead to truly significant discoveries. She says that her parents told her she could do whatever she wanted in life, and she believed them and always tried to do her best. I sincerely hope that the third woman ever to be awarded the Nobel Prize in Physics, after Marie Curie in 1903 and Maria Goeppert Mayer in 1963, will continue her important and inspiring work for many years to come.

50

PAUL CORKUM

Attosecond Science: A New Field for Lasers

Saint John, NB; Ottawa, ON

Paul Corkum, developer of attosecond science, a new field in laser science, was born in 1943, in Saint John, New Brunswick. As a young boy, Paul dreamed of working on boats. After all, his father was a tugboat captain. But as he got older, Paul changed his mind. His father was from Nova Scotia, and all the Corkums in Lunenburg and LaHave are distant relatives. His mother was born in Saint John, one of eight children (seven of whom were girls). Her family believed in education and all eight children attended university, as would Paul, his siblings, and all his cousins. This meant that Paul had a large, extended, and well-educated family.

Paul's father passed away when Paul was only thirteen and his youngest brother was four. His elementary school teacher told his mother that he was an average sort of student and "not to expect much from Paul." But by the time he was in high school, he had developed a passion for math and physics and says he owes his career to a teacher who taught "in a good way. He believed you could prove everything you could understand mathematically." Paul does not know if anyone else owes their career to that teacher, but years later, when Paul received a prestigious Killam Prize, granted by the Canada Council for excellence in scholarship, he invited this teacher to the ceremony, but learned that he had passed away. Paul acknowledged his positive influence in his acceptance speech and sent a copy to the teacher's widow and children.

Throughout his school years Paul worked at his aunt and uncle's bakery, making deliveries for them. They would give him leftovers to bring home. (He was particularly fond of the lemon pie.) The entire extended family helped Paul considerably. An aunt and his uncle Barney, who was like a father to Paul, lived just across the street. When he was in seventh grade, Paul got a paper route, which he kept for many years. It was a good route. There were many homes and people were kind to him, giving him nice tips. Then he got a job at the *Telegraph-Journal*, stuffing sections of the paper that came from elsewhere in the local publication. Every Friday he went to the newspaper office at 2 a.m. to stuff the other sections in the local paper. Sometimes the police would ask him what he was doing out at that hour and give him a ride. This gave him bragging rights among the other children at school. Then in the second summer of high school he worked on the Grand Manan ferry. Grand Manan Island is a popular tourist destination in the Bay of Fundy. From the deck of the ferry, you can see humpback, minke, finback, and sometimes even right whales, and of course, porpoises. In subsequent summers, Paul was also a deckhand on a tugboat.

When he finished high school in 1960, he left Saint John for Acadia University in Wolfville, Nova Scotia, where he studied general engineering, followed by two years at Nova Scotia Technical University (now affiliated with Dalhousie University) in Halifax. Acadia was a small university with only four faculty members in physics. It was located across the Bay of Fundy from Paul's home. He could just about see Saint John on a clear day, and when the wind was right he could smell the pulp mills. Cliffs overlooked the water, and fields were surrounded by dikes built by the early Acadian settlers, who became engineers out of necessity in the eighteenth century. Paul remembers Graeme McGarvey, who taught at Acadia after the Second World War and was head of the physics department. McGarvey had gotten research contracts in fluids with the US Navy and was able to employ students in his lab, including Paul, who worked for him from his first year at university. Paul developed films, made holographs of drops falling, and together they published a paper in *Nature* about the wake of displaced air behind the bubbles (similar to the wake behind a boat moving through water). He studied vectors, tensors, and calculus with Professor McGarvey, who gave him extra classes during the summer. Paul notes that being in a small school can be tremendously helpful, as there are unique opportunities, including a first co-publication in a prestigious journal.

Professor McGarvey thought Paul should go to graduate school and suggested that a midsized university with a strong commitment to research would be a good place for a shy, young scholar who was very talented in math and physics. Paul was accepted at Lehigh University in Bethlehem, Pennsylvania. He got married on graduation from Acadia in 1965 and left for Bethlehem with his wife, who became a student at Moravian College (since Lehigh was an all-male school). Paul did a master's in engineering and physics, and his research project was in fluid dynamics.

For his doctorate in theoretical physics, which he also earned from Lehigh, he worked on transport inside a plasma (ionized gas) placed in a magnetic field, measuring the heat and charge to learn what happens if you have an additional charge. His advisor, who was also head of the department, supervised Paul at a distance. By contrast, Paul is today more engaged and involved with his own students, and he now looks at his early work as "theory," and sees himself as more of an "experimentalist."

When he completed his PhD in 1972, Paul returned to Canada and the National Research Council (NRC), where he learned how to conduct experiments. When he was seeking a job, he first sent out a standardized letter to many places. He learned from the lack of response and changed tactics by reading the papers of the people in whose labs he would like to work and referring to them in his letters. He landed an interview with the laser and plasma group at NRC. During the interview, he was asked what made him think he could conduct experiments, and he responded, "I took my car apart and reassembled it, and now it's working." The car was a Falcon convertible, which he had purchased secondhand, with 250,000 miles on it, and smoke that poured out the back. He couldn't afford a new car, so he read a book, lined up the parts, fixed the ones that were broken, and put the car back together. He says he did this "with the arrogance of youth," but the car worked perfectly. And he got the job.

He learned about lasers at the NRC—how they amplify light, how to align them, and how to intensify the light with brief pulses. He compares what they were doing to work at the Institut national de recherche scientifique (INRS) in Québec, which has a very large laser system where they make brief flashes of X-rays to do research in agriculture.

Paul's research focuses on making a laser-like source of X-rays by using identical photons that enable him to probe things rapidly and more accurately. He looks for the ultimate time response for molecules. This has been one of the most important developments in X-ray laser research. Paul's model of ionization of atoms led to a new approach to making X-ray lasers. He now looks at solids, molecules, and atoms, and the fastest responses, showing how each electron is related to another and how they influence each other. This interactive relationship produces some of the quantum properties of solids. Using very short laser pulses, electrons can be made to move around circuits. Today, fast electronics are still relatively slow (nano-pico seconds), and Paul wondered if he could harness this technology to make them even more rapid. In 2001, he demonstrated, with colleagues from Vienna, laser pulses less than one femtosecond (a quadrillionth of a second) that are used for generating higher harmonics as a type of laser-tunneling microscope. He also speculated that the photovoltaic systems could also become more efficient.

Lasers lend themselves to interdisciplinary work involving physics, chemistry, and photobiology and, working with a multidisciplinary group, Paul came across applications such as imaging. In the 1980s, they discovered the first ultra-fast phenomenon. When he understood how to make nanosecond pulses, he thought the discovery was important, but perhaps didn't know exactly how important it would be. The discovery came about with Paul realizing that he could make harmonics with light producing unique sounds like a piano. The light waves are like sound waves. What was not known was that you could create even higher harmonics. Paul understood how this occurred. He says, take an atom with electrons moving around it. The electron is pulled by the force of the laser and breaks apart from the atom and then returns to it, creating a different kind of atom. It is, Paul says, like going to the Bay of Fundy between Nova Scotia and New Brunswick. Imagine that the atom is a rock, and the electron is flowing in the water around the rock. The wave (laser beam) goes up and down, crashing on the rock; the electron then gets pulled away by the tide and later returns with the next tide to rejoin the rock. If the rock is one atom and every rock along the shore represents another atom doing the same thing and interacting, you can see the process being amplified. This is the much oversimplified basis for Paul's 1993 paper in *Nature*. It is, he said, simple, classical physics. He thought at the time that it would not be published because it was theoretical and lacked experimental proof. Paul says you can look at light as a river or ocean with waves. He adds with extreme modesty that physics is built on just a few basic ideas, "which are complicated by mathematics." Paul's electron recollision model became the basis for the generation of attosecond pulses from lasers. These rapid pulses enabled the use of lasers to develop the fiber-optic communications we employ today. They have also been used in the cameras on cell phones, in medical surgery, and in scanning. The applications, combined with other developments, such as Donna Strickland's expansion and pulse techniques, hold vast potential.

Paul says that physics is a "specialized subject that is not the kind of thing you can share with your next-door neighbour or maybe even your wife." This is why physicists have friends around the world. They meet only once in a while and share their knowledge. They also contribute in a very personal way to international understanding. He recalls being at a conference in Moscow before the Soviet Union broke up. He was invited to the home of a Russian scientist whose twenty-one-year-old daughter had a baby with serious medical problems. Paul and his wife hosted the daughter in Canada for four months so that the baby could receive medical attention. While the baby was being treated, they asked the mother what she would like to do. She said she wanted to take an accounting course in a community college. She completed three courses by correspondence and is now one of the few people in that area of the world to know Western accounting principles. The baby successfully came through an operation, and today she is married and has a two-year-old child of her own.

"You get to know people personally and they become real people, not just ones you read about in the newspaper," Paul says.

One of the most important developers of X-ray laser technology, Paul is currently the University of Ottawa National Research Council Chair in attosecond physics. He says it is an interesting time for physics in Canada. There are many very strong researchers across the country, and all have very good, promising students. He tells his students to follow their interests and to look for unsolved problems. He reminds them not to repeat what has already been done, unless of course what has been done is wrong. For example, there were researchers in Ohio who had views on how solids work in high harmonics. Paul and one of his students proved otherwise. Paul has won dozens of important awards and medals, including the Einstein Award, the Isaac Newton Medal and Prize, the Wolf Prize, the SPIE Gold Medal, the Harvey Prize, and the Tory Medal.

Paul and his colleagues have been planting the seeds for discovery and innovation, while making extraordinary discoveries and highly innovative applications themselves. They remain on the cusp of new global trends through their own brilliance and through the kindness and collegiality they share. Paul does not need to try to be at the top of his field and the ideal professor. He already is.

51

STEPHEN MIHAILOV

Exploring the Potentials of Optical Fiber

Ottawa, ON

"It never ceases to amaze me," says Stephen Milhailov, musing on the potential uses of optical fiber. He has thought of and implemented a good number of innovative applications, but is certain there are many more on the horizon. "Think of smart structures, systems for aerospace, self-aware technologies, haptic sensing for robotics, biomedical sensors, seismology, and geology, to name just a few."

Stephen was born in Ottawa, Ontario, in 1964, and raised there. His father had escaped from Romania to Italy in 1949 and learned Italian (which he still speaks with the people he meets on Preston Street in Little Italy, a neighborhood in Ottawa). On immigrating to Canada, he met Stephen's mother, who was from a family that had lived for generations in Sudbury, Ontario. He was a metrologist and encouraged Stephen to study science and math. Stephen remembers the moon landing, watching *Star Trek*, and having dreams about studying astronomy. As a teen he recalls a documentary series on PBS, *Nova*, which dedicated an hour to lasers. This episode really captured his attention, and in 1986 Stephen earned an undergraduate honours degree in physics at Carleton University and went on to complete his PhD in laser physics at York University, in Toronto, in 1992. He studied how to use lasers for eye surgery, photo lithography, and for machining the plastics used in the manufacturing of computer boards. While a student he was able to work one summer at a local Canadian laser manufacturer and a second with Paul Corkum's group at the National Research Council (NRC) in Ottawa. Stephen says that, at the time, "lasers were a solution looking for a problem."

In 1992, he was awarded a postdoctoral fellowship at the University of Bordeaux, France, which was funded by the Natural Sciences and Engineering Research Council (NSERC). There, he worked with CCD cameras, which are digital cameras containing a charged, transistorized light sensor that takes in visual information and converts it to photos or videos. Following his postdoc, he worked for six months for a company outside of Oxfordshire, England, that had been founded by scientists from the Rutherford Lab. There, he learned about fiber Bragg gratings, which are optical fibers that reflect some wavelengths of light and transmit others. They can also be used as sensors. Since then, Stephen has worked in this field.

When he returned to Canada in 1994, he was briefly employed in his father's company, which calibrated weights for standardization and was bought by Fisher Scientific. Those were, Stephen muses, "the heady days of the tech boom. Jobs were easy to find." He also worked for JDS Uniphase Corporation for two years, developing the fiber Bragg grating. He moved on to the Communications Research Center and worked with Dr. Ken Hill's group as they invented the technology that would increase the bandwidth of networks, moving them from copper to semiconductors that would send optical signals on fiber. They discovered that they could link optical fiber with different colours and different channels, expanding by two orders of magnitude the capacity of the system. You could have a different colour for each location. For example, red could be Montréal, and blue, Vancouver, and send the wires along the same trunk, inserting and removing colours along the way as required. By using a high-powered laser, they could manufacture a structure within the fiber. IBM recognized this innovation as one of four technologies that are key to fiber optics.

Then, in the early 2000s, Paul Corkum was developing a new kind of laser that would be ultra-fast with extremely short, attosecond pulse durations (these are one-quintillionth of a second, or ten to the minus eighteenth seconds). Donna Strickland was also working on this topic, for which she later received the Nobel Prize. Stephen was fortunate to obtain one of these extremely powerful laser machining tools to advance his research in his own lab.

Stephen modified the optic fiber by using Bragg grating and femtotechnology, which is the manipulation of matter on an even smaller scale than nanotechnology. His innovation was figuring out how to make fine filaments of optical fiber from any kind of transparent material, including glass, expanding applications to reflect the environment and temperature. He also learned that fine filament optical fiber can be tied, suspend weight, and measure strain. The thin fiber was stable to the melting point of glass and could also be used as a sensor. He learned that he could make optical fibers from other transparent materials using techniques involving nonlinear optics to transform the material rather than photochemicals. The fibers could be used to monitor the exhaust from gas turbine engines and coal gasification generators. They can facilitate the fabrication of fiber lasers. Using Bragg grating as a mirror to create high-powered laser readings, the system could also be employed for directive energy.

Stephen enjoys working with industry and seeing the application of his ideas in modifying optical fiber, making optical filters and fibers that can monitor the exhaust of gas turbine engines, for example. He prefers hands-on work in his lab to purely theoretical contemplation. His next project is a patented approach for harsh environment sensing that will indicate extreme temperature changes, which can affect the way machines work and save lives. It could be used in nuclear reactors, jet engines, and for the natural environment. It might also be imbedded in the special clothing worn by firefighters, who can be inadvertently overcome by bodily heat within their protective gear. He is working on the use of hydrocarbon coal to monitor oil leakage in pipelines, and on the gasification of coal to produce hydrogen and carbon dioxide, which can be safely sequestered (and not released into the atmosphere). He is developing sensors for the green economy and the electrification of vehicles, particularly to regulate the speed of the charge, which, if too rapid, can create thermal runaway (like the cell phones that caught fire, electric vehicles can overheat and catch fire). He is also modelling how batteries will perform when charging or discharging.

Stephen's advice to young people is "If you want to do research, you must know math." He says you need to gauge what will intrigue you and stimulate the feeling of wonder so you will never be bored. One should be open-minded and question things. Stephen was determined to be a physicist because he wanted to know why and how things worked. He suggests getting hands-on experience in the subject you like. He thinks it is important to develop an appreciation for a range of subjects, including the arts. Developing spatial and audio acuity through an appreciation of music and sculpture can contribute to an awareness of optical frequencies. Painting can be meticulous or free-flowing and can also help in recognizing certain patterns and phenomena. Seeing patterns and being able to make corrections will allow you to innovate.

Stephen's work improves our environment and our lives through very significant and impressive innovations to technology. The study of science taught Stephen how to question, think, and test ideas, but it was his employment in companies that helped him imagine practical applications, and his work at the NRC that allowed him the freedom to discover and innovate. Across the country and around the world, we benefit from the impact of his work.

52

CHRISTIAN MAROIS

Photographing Planets and Stars

Vancouver/Victoria, BC

Born in 1974 in Baie-Comeau, Québec, Christian Marois grew up in Sakami and Laval, Québec. His parents loved education and his mother had a passion for astronomy. She showed him a partial solar eclipse in a shoebox, and ever since he loved astronomy. At the age of five, he was asking questions like "Why can we see the moon in the daylight?" He closely followed the *Voyager* space missions. He read books and magazines on astronomy. His friends called him "Astro," and yes, he even watched *Star Wars* and *Star Trek*.

Christian's interest became more focused over time. He specifically liked planets and stars and high-definition photos of them. When he was a teenager his father arranged for him to join an astronomy club, whose members were mostly adults. But age did not matter; he made lifelong friends there. In high school, he was permitted to take home the school's telescope, until his parents bought him his own.

He studied at the Université de Montréal, where he received his BSc in physics in 1997, his master's in astronomy in 2000, and his PhD in astronomy in 2004. As a student, he won a scholarship to the Space Telescope Science Institute in Baltimore, Maryland. He had worked diligently on his application, going to the Université de Montréal library and building the application to mimic the institute's preprints of pertinent publications. This would be his first time alone, away from home, but he was in the excellent company of fellow lovers of telescopes, stars, and planets. He also worked a summer for a professor at the Université de Montréal, where he took images of globular clusters at the Mont-Mégantic Observatory in Québec.

His advisor hired him to build a camera, making Christian one of the first researchers in the world to have worked on making images of exoplanets, the planets that are outside our solar system. Indeed, he took the first image of an exoplanet orbiting another star. To do this he needed very large telescopes and strong contrasts. He used adaptive optics to create better pictures and image processing techniques to reduce the telescope's diffraction. Previously, contrasts were weak because they only used two colours and because the instrument rotated to track the sky motion, distorting the images, and nobody had been able to correct the problem. His idea, which occurred to him during a lecture on the Hubble telescope in Hawaii, was to halt the instrument's rotation and stabilize it. It was a very simple but powerful idea. Christian created a filter that analyzed images that are captured every thirty seconds for an hour. Then he put together the software that would allow for the discovery of exoplanets, even if they are much fainter than the noise in any of the images.

While doing a postdoc, from 2004 to 2008, Christian first did a survey of eighty stars with his collaborators using the Gemini Observatory in Hawaii, but they did not find any exoplanets. He thought perhaps he was looking in the wrong area in space and should look at stars that are more massive than the sun. Then, with his colleagues, he was able to locate and photograph a new exoplanet system, HR 8799, using the Keck and Gemini North telescopes. He says he "finally succeeded in capturing the first true image of the entire solar system," and characterized this achievement as a "giant step in the quest to discover Earth-like exoplanets orbiting around stars."

Having spent three years at the Lawrence Livermore National Laboratory in California, where he liked the climate but learned how much he preferred Canada's way of organizing research and careers in science, he decided to return to Canada in 2008, where he could focus more on his research at the National Research Council in Vancouver and the University of Victoria, where he is an adjunct professor.

Christian does not lack problems to solve. For example, the exoplanets that we can currently photograph are similar to Jupiter, but much younger, only tens of millions of years in age, compared to the 4.5 billion years of our solar system. These young Jupiter-like exoplanets are still very hot and bright from their recent formation. The star light is so bright that it prevents the detection of lower-mass planets. New technological innovations are needed to achieve higher contrast and better remove the starlight from the images. On the other hand, for imaging Earth-like exoplanets, the light from the stars is insufficient to properly calibrate instruments to achieve the required sensitivity. Christian and his students have come up with a novel solution: Why not send up into space a few small satellites with lasers? They could be positioned near the star and be used as "extremely bright artificial stars" to calibrate, with exquisite precision, exoplanet imaging instruments. It would be like creating a photo studio in space. It is not yet known if this idea will work, but if it is feasible it would dramatically speed up the search for life outside our solar system.

Christian's photos can help us find out how the Earth was formed. We have much to discover about the diversity of planets, the paths they take, their different stages of evolution. Some might be in an ice age, others might be home to dinosaurs, and yet others, like the Earth, thriving with many species. We can learn about the origin of all forms of life and what factors impact evolution, such as the role of tides in our oceans or how Jupiter might have helped clean up the small asteroids in the solar system and limited large-mass extinction impacts, allowing life to flourish on Earth. We may even find other civilizations. There are so many possibilities and so much to learn.

Christian believes that in our lifetime we will have the first images of planets that resemble the Earth. We will possibly find traces of life. He thinks there will one day be a system of telescopes capable of seeing the surfaces of continents on other planets. The James Webb telescope's pictures that made the news in 2022 were a great step forward, but new technologies must be introduced in the next mission, so that pictures of rocky Earth-like exoplanets can be taken. In the next decades, we will be able to capture images of vegetation and continents, and all that exists on these planets.

Christian never had a single doubt about what he wanted to do in life, and he is still passionate, enthusiastic, and ready to jump out of bed in the middle of the night to peer at the sky. His advice to youth is "Follow your passion. Arm yourself with patience. Sometimes it may seem as if you are moving two steps ahead and two steps back. But remember, each step forward is progress toward your goal."

PART NINE

Social Innovation for a Better World

We celebrate the latest app on our cell phones, the new, inexpensive sequencer, the largest synchrotron, or perhaps the synthesizer that provides a new experience in surround sound. While we also applaud ways to map the world, preserve and celebrate language and culture, improve the delivery of health care, seek to be a more inclusive and caring society, and promulgate rational and reasonable laws, we may not think of these activities as innovations. They are, however, extremely innovative, and the changes they effect are profound and meaningful. As we seek to understand the universe, we must attempt to improve life and the possibility of life on our own planet.

The individuals in this section base their worldviews on their international experience. Their personal stories validate their care and concern for the world and for their communities. They combine disciplines such as art and data science, literature and legislation, population studies and health, economic modeling and the study of social and cultural values.

Inspired and inspiring, these innovators continually listen and learn from others. Whether they seek to include and offer possibilities and hope to populations without access to health care and education or gain inspiration from young children's views of a more perfect world, their accomplishments result in significant change with potentially extraordinary impact.

53

BARTHA MARIA KNOPPERS

Creating Space for Passionate and Rational Dialogue

Montréal, QC

Personality of the Year, Canada's Nation Builder, the Academy of Great Montrealers, Officer of the Order of Canada and of Quebec. These are but a few of the awards won by Bartha Maria Knoppers, who arrived in Canada from the Netherlands in the second wave of immigration after the Second World War. She recalls arriving on a large ship and celebrating her fourth birthday on board. It was the biggest birthday party she had ever experienced and certainly presaged the celebratory applause she regularly receives today for her extraordinary contributions.

Her parents established a church and school in Edmonton. Her mother had been working on a degree in divinity in Amsterdam prior to the Occupation, and she showed her seven children that learning was a pleasure, a personal and intellectual source of joy, encouraging them to read. Both parents were highly supportive of studies, and only when Bartha Maria decided to pursue a doctorate in law instead of literature did she detect the tiniest twinge of regret from her mother, who would have rather liked to have a poet in the family.

Every year, the entire family travelled and camped across North America with a Coleman stove. The children learned to appreciate the differences among regions, while the parents enjoyed both the time with them and the break from pastoral duties.

At the age of seventeen, Bartha Maria went to Calvin College in Michigan for two years, and then transferred to McMaster University in Hamilton, Ontario, where she completed a BA with honours in French and English. She worked as a waitress in the summers to help finance her studies and learned a great deal about human nature. She saw both complexity and great kindness. Then, at the University of Alberta, she did her master's degree in poetry and Caribbean literature. Her PhD thesis at the University of Alberta in Edmonton would be about how colonized nations can create their own literature and how national literatures can develop. In 1976 she ended up changing course, going on instead to spend another seven years studying law in Québec, France, and the United Kingdom, where she met some truly "beautiful thinkers." In every field of law and science, issues concerning reproductive rights were present and became the topic for Bartha Maria's doctoral thesis in comparative medical law, which was timely, since it was during that period that the first in vitro baby was born in England. Looking back on her years at university, Bartha Maria reflects that study affords us the opportunity to see, experience, and understand everything. If she had to do it over, she would gladly do so, and this, despite the burden of student loans. In her early, pre-university years, she had avoided every science course that was not mandatory. She does regret that while she studied mathematics, physics, and computer science, she did not study biology or chemistry. However, she certainly has made up for this gap. Today she notes she works in the department of human genetics, faculty of medicine, not the faculty of law, and the first article she published was in 1978 on artificial insemination!

In secondary schools across Canada and in CEGEPs in Québec, students today often study prelaw and pre-medicine. Bartha Maria thinks that this is a narrowing of futures in every sense of the word, both professional and human. She advises youth to enroll in the most comprehensive curriculum. She says, "You can always change your mind." Reading is her biggest pleasure. She absolutely devours books and often buys them just because of their titles. She cites *The Colony of Unrequited Dreams* by Wayne Johnston as one example. The title draws you in, especially when your work has something to do with dream fulfilment in an imperfect world, with finding the common ground to discuss difficult issues and creating a context in which these matters can be discussed.

Bartha Maria's law professor at McGill recommended that she enroll in the program in social rights at the Sorbonne in Paris, thus catapulting her into international studies. She completed her doctorate at the Sorbonne, followed by a postdoctoral program and a diploma in legal studies at Trinity College Cambridge, where she studied prenatal diagnosis and legal liability. The contrasts between Paris and Cambridge were immediately visible, starting with the gowns you wore and the formality of the British education system. Her first child was born in England, and she observed through experience the excellent health system and the commonly held belief in universal health care, which included access to biotechnologies when safe. Bartha Maria is most proud of her work on commissions both in Canada and abroad. She notes her participation in the recent international commission on human gene editing, and when asked the hardest thing she ever did, she jokes, "You mean, besides raising children?" However, she quickly points to her role on the Canadian Royal Commission on New Reproductive Technologies (1990–93). The work was important. Up to 15 percent of the population cannot have children. She says we all need a minimum of science to understand the medical conditions involved in infertility, and we need to talk about the broader issues that are all based in the consideration of rights. The diatribes launched against the commissioners were very political. The questions being studied were oversimplified by people who resorted to the lowest possible denominator of public discourse. It was "hard, hot, and disheartening." She observes that, these days, politicians live very lonely, all-consuming, and extremely difficult lives, which she experienced secondhand when her husband spent twelve years in politics. She loves teaching and her own research center, but rues the tendency in some areas of the world to limit what can be taught and researched.

Part of the problem today is the anonymity afforded by the internet. Individuals can hide in the morass of communications. People have lost the ability to listen to one another. However, Bartha Maria believes that this will eventually change, and that the gratuitous intellectual and physical violence will end. People are, after all, "fundamentally kind and have common sense." The tone will change when people go beyond the stuff of tabloids and are themselves engaged in service to others.

She evokes issues raised by the 3D printing of cells and organs, gene therapies, and the necessity to remember the suffering of individuals as well as the common heritage of humanity. When Christiaan Barnard did the first heart transplant, it was criticized, as the heart was seen by some as the soul that identified an individual spiritually. Today, the brain and the genetic "code" are the "final frontier." We must not base policies on just a few examples or be stopped by the risk of failure. Rather we must open issues and discuss possibilities for misuse or abuse, but always return to shared values on human rights. We must have the goal of saving lives or at least diminishing suffering.

Bartha Maria is currently framing the universal human right to benefit from science to include open science, information technologies, and policies. This relocates the discussion away from the more adversarial, balkanized ways of viewing ethics and law to a more universal ground. She considers policy as a form of infrastructure. Data sharing can be seen as democratic. We need to standardize data codes and entries to promote interoperability around the world. Genetic research, social media, and uses of information technology should reflect a modern, heterogeneous Canada, and she sees a need for transformation of discourse from the binary to the universal for the benefit of all.

Her goal for the future is the creation of a series of prospective and principled charters that will be expressions of, and visions for, improving scientific integrity for humanity. She believes we all have a duty to serve others.

If Bartha Maria were a scientist, she would probably have invented ways to create vaccines and remedies and bring them to the needy around the world. If she had been a writer, she would probably have written a utopia, a tale of life in a perfect world. She *is* a brilliant, creative legal scholar who seeks universal principles to establish common ground. She does not seek to invent a cure for the ills of society; she innovates and creates an environment, an unimpeachable space in, and from which, reason and rational thought may build a solid basis for agreement. She does not participate in debates where the loudest voice triumphs. She steps aside and constructs an intellectual, rational place, an island of innovation, where the finest qualities of humanity are regarded with respect, where hope reigns and perhaps even dreams will one day be requited.

54

ANNA TRIANDAFYLLIDOU

Innovating the Dialogue on Migration

Toronto, ON

Anna Triandafyllidou provocatively says that the fall of the Berlin Wall was perhaps the most significant event in our time. This realization caused her to turn from the study of political parties to the study of migration, management policies, multiculturalism, migrant domestic workers, highly skilled migrants, education, and political participation. She questions how it is that entire countries move from being xenophiles to xenophobes, a topic of particular importance in the world today. She states that migration is as much about the people who stay as those who leave. We tend to focus on the places people left and those to which they came. She thinks about the spaces in between. These are refugee camps and temporary dwellings. They are also the transnational spaces we all inhabit thanks to education, the media, and the mélange of cultures and traditions we inherit and share.

Anna's parents and grandparents were all lawyers, and Anna and her brother were expected to go to university and possibly even follow the family's tradition. As a young girl, born and growing up in Athens, Greece, Anna had an inspirational professor of Modern Greek language and history. She had a "very genuine way of connecting with students, and when she left the school, her classes gave her a standing ovation." Anna pursued her studies in sociology, as she wanted to understand how society "works."

She studied French and English while in secondary school and also learned Italian and German during her university studies. Upon completing her BA at the Panteion University in Athens, she wanted to go abroad and did some research on courses and scholarships available, visited several embassies, and saw a poster about the European University Institute in Florence, Italy, where she was admitted directly into the PhD program at the tender age of eighteen. Although she was comfortable speaking Italian, she was made to feel as if she would never quite be accepted and would always be a "foreigner." This attracted her interest in social psychology and identity studies. She also worked various jobs during her studies, as a legal typist, operator at a call centre, tour guide, and English-language teacher. She thinks it is important for young people to work in order to gain experience during their formative years.

Anna studied and worked in Greece, Italy, Belgium, Great Britain, and the United States, before choosing Canada, where she was offered, in 2019, a Canada Research Excellence Chair in migration and immigration at Toronto Metropolitan University (formerly Ryerson). She has personally experienced the pride in place and country that children learn in school and at the knees of their grandparents. She has also witnessed the fall of the Berlin Wall and all it symbolized, helping fashion the Europe that the philosophers of the eighteenth century only imagined. She has seen the tides of migration caused by natural disasters, strife, and poverty and has understood how migrants and immigrants feel, as well as how those who have worked hard to establish strong economies and vibrant traditions feel when it seems that all they have worked for is at risk. Anna's own family, with her children having been born in and lived in different countries, is perhaps the illustration of the family of the next century. In the past, when populations became transitory, moving for work, the nuclear family was created out of necessity and society developed around that unit to support life. If today, families are multinational and take international transport as easily as local transit, they will seek different supports for their lives. Languages will become increasingly important, as businesses and families both need to communicate to succeed.

With her personal experience and considerable expertise, Anna is widely published and called upon for advice. Her innovative contribution to society is to open difficult discussions, recognizing the place of the politics of cultural and national identities, but then centering the debate in a rational framework of history, logical analysis, and a clear statement of needs and goals. She has created a space for the consideration of very difficult questions in the university classroom, the research community, and the world at large, as she is frequently invited to offer both academic lectures and practical advice to governments around the world. She has included in this space ethical and moral values that transcend individual cultures to enable a kind of global humanitarianism.

Anna has become convinced that while migrants cannot be guaranteed the right to immigration, migrants can be offered the humanitarian right to asylum. She says, "Some nationalist scholars do not think of the broader picture and get weighed down in the 'nitty-gritty' of policy, forgetting why this is happening, why countries that appeared to be models are now suffering backlash." We need to examine the evidence carefully, *before* making policy, *not after*. We need to look at the bigger picture, which includes history as well as the technology that has changed our lives. We must ask what is work and how our conception of what is "good work" is changing. It is essential that we consider how we will deal with secularism and diversity. This exemplifies the way she opens topics for dialogue and focusses on core values.

Anna sees a future where technology will transform every aspect of our lives. "Technologies will impact migration. We may all become digital nomads. We will be able to work from anywhere, but will still want to migrate, which is different from migrating to find work. Big data and digital surveillance will be even more important issues. What data should we generate and how will we use it? How can we train artificial intelligence engineers to be EDI-minded (conscious of the principles of equity, diversity, and inclusion)? Social media will play an important role, and we must decide if it is for polarization or democratization."

She imagines a future where we mobilize for climate change, where we create governance for migration that is compassionate and promotes inclusion rather than exclusion. The world, she says, will be ever more mobile, yet people need anchors.

Anna says she is Greek by origin, European by heart, and Canadian by values. Those values are inclusion, equity, and diversity. Anna says that the perfect is the enemy of the good, but in our imperfect world, she is pleased to have chosen Canada, which is, she concludes, a very fine country. "Canada is," she says, "a good place, not perfect, but a bit like a marriage, you have to keep working on it." Anna's thoughtful, dispassionate, global views and her expertise are most important in our world that has created the borders across which displaced populations are moving in pursuit of refuge, new lives, and the ability to support themselves. We are fortunate to welcome global citizens like Anna to enrich our knowledge and provide the framework for healthy debate and lead international consideration of difficult topics with reason, based on evidence, compassion, and strong ethical values. Logical, rational thinking, and humanitarian concern that includes ethics have been taught for centuries. Innovatively applying them today to issues, such as migration, that are the subject of passionate debate and the cause of violence around the world requires great wisdom.

55

KATRINA MILANEY

Listening and Solving Problems with the Community

Calgary, AB

Katrina Milaney, born and raised in Calgary, has a natural curiosity about the world that would have taken her on a path of scientific research in biology or earth science had the experience of life not pushed her in another direction. She says that all the social issues she studied while writing her PhD dissertation on the social determinants of health were part of her own difficult story. From a young age, she knew there was injustice and thought and felt deeply about the way society deals with difficult situations. At the age of twelve, she already called herself a feminist, surely presaging her later studies and work.

She began her university studies at the University of Calgary in psychology and then moved to sociology. She worked for a dozen years in public institutions, including libraries, before returning to do a master's in disability studies, which combined the theoretical openness of sociology with her search for meaning and openness. She learned the language needed to give voice to her concerns around equity and the skills in critical inquiry needed to ask questions about difference such as: Who defines difference and who determines who is different? Her mentor observed that Katrina was a critical social theorist. This led Katrina to write a dissertation on women in prison. Continuing along the same interdisciplinary path, Katrina now finds herself, a social scientist, located in the faculty of medicine where, she admits, she at first felt like "a fish out of water," but soon found her place thanks to wonderful colleagues.

Katrina's entire approach to academic research is innovative. Before beginning her graduate studies and while she was working, Katrina often helped academic researchers frame their studies. She collected data for them and never heard from them again as they published and gained accolades for their work. Now that she is a researcher herself, she says everything she does is really grounded in the community and this is where her innovation comes to the fore. The community, not the Ivory Tower, is where both the issues and the answers reside. Every project she undertakes is done in full partnership. She co-designs projects with experts in the community. A recent success brought a grant of $3.5 million to community agencies. The research and work that she will respectfully help guide will contribute to positive change.

She offers an example of a project to address homelessness among Indigenous women. They have often been offered inappropriate and very poor conditions by racist landlords who felt they had the upper hand, as the women had nowhere else to go. Katrina consulted elders in the community. Should they complain to the municipality? The elders said the problem was the landlords, not the municipality, and they should work to develop a landlord network, explaining to them the structural racism in which they were participating. This is, Katrina says, just another small step toward change.

In working to alleviate the problem of homelessness, she turned the problem on its head. Rather than looking at making shelter spaces open to longer-term stays, she is working with the community to help people return home before becoming entrenched in homelessness. They try to bring families back together as soon as possible. The program is working in five cities across Canada, and there are over five hundred youth and families who have been reunited, and nearly half have remained at home. This is a shift toward prevention. Youth and families receive assistance to build good relations. If youth remain homeless, the chances are that this will affect health. People may be trapped into homelessness as a result of complex issues and need access to multiple services. Another project in which Katrina is involved maps the care system and provides information on how to navigate it.

Katrina also collaborates with a project in Victoria, BC, on low-barrier housing, focusing on harm reduction with traditional, cultural teachings, including smudging. Katrina is impressed by the spirit of hope she feels when talking with those in the program who are determined to improve their lives.

She is involved in at least a dozen projects and knows of many more. She is humble about her contributions, but her role is key. She contributes knowledge and experience, heartfelt concern and caring, and national connections and scope to place her projects in perspective. She says that she feels obligated to pursue her work with the community. She knows she can have a positive impact and, "honestly, there is nothing more fulfilling. There is meaning in my work." She begins by reducing barriers and then she steps back and muses that while the health-care system may work fairly well for average people, it does not work for those who really need it. Think of victims of family violence, people with handicaps or addictions, those who are ill at ease in office or hospital settings. If we design the system for those most in need, it will work for all.

Katrina dreams of the day when universities will be recognized as leaders in innovative actions. She imagines the day when, for example, people will no longer be classified as "vulnerable." To say that senior women are vulnerable is to ignore the fact that many are strong leaders. The problem is not that people may be vulnerable, but that the system is broken, and policies are not appropriate. It is a matter of structural vulnerability. She says she wants "universities to be fully engaged with their communities and grappling with complex and truly wicked problems."

Katrina recognizes the work of her mentors, including Bill Ghali, and her peers and colleagues both in the university and the community. She states, "If you want to do this kind of work, you must have patience, openness, be able to step outside yourself, listen to differing opinions and find common ground." Katrina admits that the work can take time and be difficult; however, "it is *so* worth it at the end of the day." When she sees a family that obtained safe housing where the children can just enjoy being children, it is extraordinarily satisfying. She says, "You need to reflect on why you are here and care about people. You must focus on impact." She is "hopeful—always—and perhaps without reason, when we look at the state of the world. We seem to go two steps ahead and two steps back. But I am still hopeful." She reflects that she could live and work anywhere in the world, but she is where she wants to be, at home in Calgary, in the community and at the university.

Innovation is about taking consistent, persistent problems and looking at them through an entirely different lens, and the focus must be on people. Recently, when Katrina was speaking to a mother, she asked her, "What do you need? What would change life for you?" The woman responded that she had two children who had been assigned to two different foster families because she was deemed not to be a good mother. Each of the foster homes was given $1,400 a month to care for the child. The mother asked if they could not instead pay a social worker to come to her home and teach her to be the good mother she wanted to be. Katrina muses on the value of keeping families together in homes with support, and she worries about homelessness and violence and the fact that people need someone to listen to them. Katrina has most certainly a gift for innovative thinking, and the work she does with such great care and respect is a gift to her community.

56

MAYDIANNE ANDRADE

A Model Scientist Opening Doors for Black Scientists

Toronto, ON

How do you turn fear into fascination? It is hard to believe that one of the world's experts in spiders started off just like many of us, not particularly fond of those web-spinning arachnids.

But let us begin in Burnaby, British Columbia, where Maydianne arrived at the age of three with her parents, who came to Canada from Jamaica. They lived very near Simon Fraser University. Indeed, it was only one short bus trip up the hill, and Maydianne credits her parents with having chosen the location strategically, as she never doubted that she would one day attend university. Her parents were excited about anything she brought home—even spiders! They told her and her brothers that they did not care what they chose to do as long as they did it well and twice as well as anyone who was not Black.

Maydianne confesses to having been a bit of a nerd. At the age of three, she begged to go to school with her older brother and received early admission to kindergarten. She knew she would like to study science ever since she was in the sixth grade. She recalls watching *The Nature of Things*, with David Suzuki, never imagining that she would, in 2020, be featured on the program. She also remembers a science teacher in grade seven who challenged students to think. One rainy day, he asked them to design an experiment to demonstrate without going outside that it was raining down from the sky and not up from the ground. The teacher was so engaging that Maydianne even contemplated the possibility that one day she might become an educator. Then, in high school, she discovered she was good at biology and thought about becoming a medical doctor.

She worked in the summers and says she particularly loved her first job operating a ride at the Canadian National Exhibition (CNE), as she was able to interact with children, an experience she repeated as camp counsellor and as a salesperson in a children's clothing store, where she became an expert in fashions for the very young. Simon Fraser University awarded a scholarship to the top student in each high school in British Columbia, and Maydianne received that honour for her school. She registered for the co-op program and worked the first summer at Agriculture Canada, where she learned about the efficacy of organic farming methods and spent the following summer working in a lab. Subsequently, she received several scholarships from the Natural Sciences and Engineering Research Council (NSERC) and, while preparing for graduate study, wrote a paper that was later published on the damage to vegetation by flies.

When Maydianne began work on her master's degree she started with vertebrates and moved to fish and finally decided she really liked insects, which could be studied at scale and observed in the field without altering their habitat. In her animal behaviour class, students were required to write a practice proposal for a research grant. Inspired by reading, she wrote a sample application about sexual cannibalism among insects. Her supervisor suggested she research this topic for her thesis, which ended up being titled "Mating Behaviour and Constraints on Reproductive Success in a Spider with Male Sexual Sacrifice." She had begun with the praying mantis, but moved on to study the mating habits of spiders belonging to the *Latrodectus* species. In particular, she concentrated on the redback spider, which is common in Australia and quite invasive, now being found in Japan as well. Maydianne was interested in the topic because it seemed counterintuitive that a cannibalistic species would evolve and become an invasive species, rapidly growing in number, when the act of procreation led to the death of the male spider, which would be consumed by the female. Maydianne has travelled and done research in Australia several times, and she spent long nights in an agricultural station studying the habits of this nocturnal arachnid. There are at least thirty species of black widows, but data is only available for four or five of them. Her master's thesis led to an article published in *Science.* Today, Maydianne still has things she would like to learn from these spiders. She would like to know about the entire genus, or family of organisms, to be able to learn about ecological and genetic processes. And she gratefully recognizes the encouragement of her professor, who enabled her studies and supported her publications.

She went to Cornell University in Ithaca, New York, to continue her graduate work and was immediately impressed by cultural differences between Canada, Australia, and the United States, where science communications were part of the training and students were expected to fund their studies by writing grants. Following this introduction, the importance of science communications has become an important aspect of Maydianne's work and one that she has strongly promoted throughout her career. She is a sought-after speaker with local student groups as well as scholars on the international stage. She hosted a weekly podcast during the pandemic aptly titled *The New Normal.* She has appeared on *Quirks & Quarks* and *Nova ScienceNow* and been recognized as "One of the Brilliant Ten" by *Popular Science* magazine. It is unusual for scholars to gain a place on the public stage, and Maydianne's success is truly remarkable.

After completing her PhD, which combined biology, neurobiology, and behavioural sciences, she decided to return to Canada and the University of Toronto, where, in addition to being a Canada Research Excellence Chair and professor, she has also served as vice dean of equity, special advisor to the dean, and acting dean of science in a faculty with sixteen departments.

Maydianne has introduced to her campus many new programs and activities, including mentoring, cohort building for faculty, and education against unconscious bias. She poured all her energies into making change. It was backbreaking work, and she was on call 24/7. She is considered a pioneer for the bias training she brought to Toronto, and she received the Ludwig & Estelle Jus Memorial Human Rights Prize in recognition of her efforts. We often consider innovation finding a new way of doing something in the lab, adopting a new process to reduce costs and increase speed. Innovation is also creating change in society and our environment.

Maydianne has helped improve and change processes for hiring, tenure, and the evaluation of research. This is innovation on a large, social scale.

The pandemic "changed us all, and it is time to come together to make a positive, long-term effect on the system," she says. "Discrimination is still a fact. Black students are, even today, pushed out of academic courses." Recently, Maydianne heard an elementary school teacher state, "Blacks cannot do math." She muses, "If we look at social constructs like scientific problems to be solved, we might develop recipes for improving the environment. We could create mechanisms to change the future and mitigate change in our human climate, as well as climate change for our planet." One mechanism Maydianne created is the Black Scientists Network of which she was a co-founder and inaugural president.

"Science provides us the methodology to find answers. Innovation lies," Maydianne says, "in integrating understanding from different angles, from study and research, and from contact with others to improve on definitions and methodologies. Different fields of inquiry, including psychology, enrich the learning process and the possibility for innovation. Human beings are like sophisticated robots. If we understood the brain completely, we could answer all the questions that lie before us. Our brain is the most powerful tool we have." She encourages students to expand their horizons and study at more than one university and travel when possible.

Maydianne, the ecologist, is truly innovative in her research, looking at a small, deadly spider with unusual behaviours, and seeing the development and spread of species—the greater evolutionary picture. Maydianne, the science communicator, is innovatively reaching out to the population at large to share scientific information and methodical ways of thinking and understanding the mass of information that surrounds us. Maydianne, the innovative advocate for equity, diversity, and inclusion, has created models for change within universities. She takes an innovative approach to every challenge she faces.

Maydianne believes in "informed speculation and creative fun." You need first to know what is already known, and this can sometimes seem like "drinking from a fire hose." Early in her career as a researcher, studying the behaviour of extremely interesting and deadly spiders, she held the camera equipped with a macro lens in one hand and wrote with the other. Today she has one of the rare macro-videography rooms in the world. She has created a rich scientific and inclusive environment for research and looks at statistical differences among complicated research findings to discover new information. This is especially important when you consider ecological and environmental shifts in the world, and the fact that, without diversity, we will never realize our true potential for innovation. But there is hope because Maydianne has brilliant ideas, boundless energy, and many students from across Canada and around the world in her lab. She is inspiring our future: the next generation.

57

JENNIFER LEASON

Improving Indigenous Maternal and Child Health Care

Calgary, AB

Jennifer Leason, a Saulteaux Métis Anishinaabe, member of the Pine Creek First Nation, is wise beyond her years, creating hope for improved maternal and child health care. She is creative beyond measure in the way she brings people and organizations to work together, in the manner she conducts research and mentors her students. Her brilliant writing and works of art contain the power of her spirit and the light of her mind. Jennifer's Saulteaux name, Kessis Sagay-Yas Egett Kwé, translates as First Shining Rays of Sunlight Woman, and it is that luminous quality that permeates her every thought and movement.

She was born in Hudson Bay, Saskatchewan. Her father was a settler of Norwegian and Ukrainian origin, and her mother was Anishinaabe. They lived in Hudson Bay until Jennifer was ten, when they moved to Dryden, Ontario, a small mill town, where her father worked as a millwright. When she was fifteen, Jennifer went to live with her sister in Red Lake, Ontario, where she had a number of jobs, including waitressing at the Chicken Shop, where she made $2.90 an hour, while going to school. She completed her secondary schooling and was told by a teacher not to bother about applying for university or for a scholarship because "you are not going anywhere." Jennifer proceeded to apply anyway and was accepted and went to the

University of Saskatchewan, where she loved biostatistics and earned her BSc in psychology and neuropsychology. She is the perfect illustration of the saying "Don't tell me I can't because I will just prove you wrong."

After completing her degree, she worked for Correctional Services at a maximum-security prison, for the Department of Social Services, for a school division as student counsellor, and for the BC Ministry of Children and Family Development (Child and Youth Health). From these jobs she saw firsthand that Indigenous people were overrepresented in these services, and she saw the inequities, barriers, and gaps.

Then she worked for the Canadian International Development Agency and was sent to India. When living in a small tribal village and working on a project to reduce the mortality rate of women in the rural areas of Jharkhand for the Ekjut foundation, she met one of the local doctors, who asked her how it was that, in India, they had eradicated tuberculosis (TB), while in a rich country, like Canada, there were TB outbreaks in Manitoba, close to the community where her mother lived. For Jennifer, this was an aha moment, which prompted her return to Canada and her application to graduate school in the field of maternal and child health. She explains simply that the "state of Indigenous women's health care is appalling." She thinks that the biggest impact we could have on population health and well-being would be to invest in women's health and fetal care. She completed her MSc and a PhD in public health epistemology, perspectives, and Indigenous studies at the University of British Columbia (Okanagan).

She was immediately offered a faculty position at the University of Calgary, where she holds the Canada Research Chair in Indigenous, Maternal and Child Wellness. She is studying heath care for women on remote reserves as well as in urban centers in Canada and the US. Her research is fundamental, but also has immediate applications. Much of her work is community-based, and she collaborates with the National Association of Certified Professional Midwives, the Amautiit Nunavut Inuit Women's Association, the Native Women's Association of Canada, the Métis National Council, the First Nations Health Authority, with communities and tribal councils, as well as directly with traditional birth workers and doulas.

Her current challenge is to access data on the provincial level to identify indicators and outcomes and to deal with the many issues inherent in the use of data. She says it is essential to ensure respectful ways to access such information. This has never been done in Ontario, for example, where she suspects that very few have ever accessed the Better Outcomes Registry & Network (BORN). She wants to show what the indicators are, publish and share them and make policy recommendations. She is including economic analyses, such as the national costs of travel to distant hospitals for birth. This is, as well, the first time the socioeconomic implications have been analyzed. Jennifer is hopeful that her work will result in better health care and access to it.

Jennifer clearly sees this as a social problem and is convinced she knows the cure. We do not have sufficient support in communities that need doctors but also midwives and doulas. The solution is more training. She recognizes, however, that knowing the solution is not enough. She has to show the cost and impact. She also knows exactly how to do this. She is collecting and analyzing the data, documenting case studies, decolonizing the narrative, and collecting photos to illustrate her work.

Jennifer wants to create a space for transdisciplinary, land-based, and experiential methodologies that will help her inspire Indigenous students to come to university and become the next generation of scholars. She says her work is about reclaiming, reconnecting, and researching Indigenous values, methods, and epistemologies. She is working to create new possibilities using traditional ways of knowing and thinking as we move along a path that leads to better women's health.

Jennifer is also an artist and says her art represents her own healing and that thinking creatively is the biggest gift we can have. When she begins to think imaginatively, a series of Anishinaabe ceremonies and protocols come to her mind. She sets her intentions and prayers, and then the visions become clear to her. She hopes that her art as well as her research will have an impact and will inspire others to think and then to think again differently.

Her work as an artist came about later in life as she applied herself to relearning her language and traditions. She wrote *Meennunyakaa/Blueberry Patch: A Saulteaux-Anishinaabe Story of Traditional Berry Picking* and *The Super Incredible Big Sister Book* for her children. She paints and makes prints, incorporating art in all she does, including statistics. She begins with an image and then integrates the story of the art to contextualize where we are.

Each picture is also a story that introduces cultural teaching and Indigenous epistemology to effect change.

Jennifer crosses disciplines, mixing art and science. Her work is at once global, national, and regional in scope, but very specific in focus. She combines teaching and research, and both are framed by her art. She says the only way to make change is to find our voice and to speak out. One of the many innovative gifts Jennifer has given the world is her program "Education for Reconciliation. Experiential Learning Exercise." It is transformational and has been recognized across Canada.

Jennifer recalls and paraphrases the wise words of Elder Jim Dumont, "You are a spark of the Greater Spirit. Everyone deserves to feel good about who they are." She adds, "Never let the darkness overcome the light. Keep going despite the difficulty of your work, no matter how dark your darkest days are. Even small steps can make the biggest change." Jennifer is inspired and inspirational, and innovation springs forth in her research, teaching, and art as she shares her light and provides us all reason to hope.

58

TOLULOPE SAJOBI

Innovative Solutions and Inclusion through Data Analysis and Example

Calgary, AB

As I stumble over the pronunciation of his name, Tolulope Sajobi smiles and says, "Please call me Tolu. The only person who ever used my full name was my dad when I did something of which he disapproved."

So let us begin by imagining Tolu growing up in Ibadan, a large city in the southwestern part of Nigeria. The second largest city in Africa, it has almost four million inhabitants. Tolu's dad was principal of a secondary school and his mother, a teacher and principal of an elementary school. Tolu reminds me that English is the official language of Nigeria, which was, like Canada, colonized by the British. Besides, his mother taught English. In Nigeria, education is considered the greatest good one can offer one's children, and he and his three siblings were able to concentrate fully on their studies.

Tolu considered possibly majoring in chemical engineering, law, or medicine at university, but ended up studying math, a subject in which he excelled. During the course of his studies, he filled in as a supply mathematics teacher for a semester in a high school, an experience that offered him an early taste for teaching. He also did an unpaid internship at a teaching hospital with the intention of learning computer science. He was assigned to travel around the region collecting information on incidences of dementia in elderly people and entering this data. Many medical students would stop by and ask for analyses when his supervisor was away. Tolu figured out how to do regression analysis and provided the answers. Seventeen years later, he was sitting on a research ethics board in Canada, and someone asked if he was familiar with the famous dementia project in Nigeria. It turned out to be the very project for which he had provided assistance. Not only had he developed his own interest in statistics, but he had also helped with what turned out to be an influential project.

Along the way he was inspired by all his professors, and in particular, his undergraduate honour's supervisor, who was often invited to international conferences. Tolu was surprised to learn from him that mathematics, data analysis, and statistics, which were not the most popular of fields in Nigeria, were important areas of research around the world. He decided he would therefore like to do graduate studies abroad and was invited to study at the University of Windsor in Ontario. He admits he was a bit lonely at the start, but was very grateful to other students who helped him find an apartment that actually had heat. While Windsor is located at the southernmost tip of Canada, its climate is definitely not tropical, and Tolu spent his first four months in an unheated unit, suffering in silence and wearing his boots indoors, convinced that Canadians must be stalwart and perhaps impervious to cold. Since then, he always tries to "pay it forward" and help other newcomers to Canada.

Tolu's graduate experience was unique in that he graduated before he completed his course work. His supervisor was busy organizing an international conference and, after the first term, gave Tolu a few problems as a kind of test prior to beginning his thesis. Tolu worked on the problems and found interesting patterns between numbers and systems and discovered an important link between number theory and systems theory. His professor was impressed, and suggested Tolu write an abstract to submit for the conference. He was invited to present his findings as a paper, which he wrote with his supervisor. This was the first time a paper was ever accepted prior to peer review, and Tolu was certainly the youngest person at the conference.

After the conference, the professor said that his results were sufficient for a thesis, and Tolu should write some papers for publication. Tolu wrote several and they formed the basis for his thesis, which was completed in August of the same year. He remained in Windsor and retroactively completed the three remaining courses for his MSc, which included applied probability, theoretical statistics, and operations research.

Tolu was then invited to the University of Saskatchewan in Saskatoon to do a PhD, where he was the first student in the newly inaugurated department of biostatistics. He won a Vanier scholarship, the first international student at the university to do so. He was assigned to look at machine learning models and statistical methods for making sense of questionnaires. For example, when looking at the quality of life, identifying which dimension (social, physical, mental health) is most important to the patient, can assist the medical practitioner in making informed interventions suited to the individual patient. Tolu's work showed that we cannot paint all people with the same brush and that it is important to integrate patient voices into how clinical decisions are made. It also pointed to the difficulty of such reporting, as impressions vary a great deal from one individual to another. The question Tolu needed to resolve was "How does one identify heterogeneity?" This was both profound and innovative, as it points to a new way to organize health care.

Another project at the University of Calgary, where he now is a professor, had Tolu helping to design clinical trials. Trials are very expensive. Was there a way to reduce the costs? His innovative solution was to embed clinical registries in the process of developing the trials. This was just a creative first step. Then, looking at the university and across Canada, he saw that the country was well positioned to lead such an initiative internationally. This project was put in place at the University of Calgary for the second-largest stroke trial in Canada. It was started just prior to COVID and used existing data from the registries and was completed by January 2022. The study was published in *Lancet* and compared different methods (surgical and pharmaceutical) to open or clear arteries in stroke victims. The study found that the drugs were more effective. Tolu is now cogitating upon ways to create a registry platform, where one might simultaneously find answers to multiple questions about different conditions requiring the intervention of medical professionals from several different fields.

Tolu says that students ask him how they can be like him. He tells them that they all have a unique opportunity to lead and must always be inspired to learn new things. He adds that his field is very special, as it gives him the opportunity to be in "everyone's backyard," where you can really change the world and effect medical innovations and thus change clinical care. He says you cannot do the same things over and over and admits that he once worked for nine months in a biotech company and was bored by the end of the first week. He considers academic research more fun and likes learning what other professors are doing, from orthopedic trials to the analysis of deep data for public health to creating models for machine learning. He says a statistical leader must be able to go far beyond analysis and understand the area, the goal, and the team to comprehend the contextual area you are being asked to design. At the same time, he needs to understand what the research is. He tells researchers to explain their projects to him as if he were their partner. Anyone can be a data analyzer, but Tolu aspires to be a leader through collaboration. He believes in "stepping up and stepping out."

He feels that his international experience has helped him step out of the box himself. He recalls a meeting in 2018 with a group that had been collecting patient-reported outcomes for the last twenty-five years. They asked if there were a way to make the collection and use of their data better. He raised patient patterns and discussions with patients to determine not only if the methods were appropriate but if the patients themselves were well served. Tolu is a statistician who truly cares. He says that behind every data point lie interactions with patients and their families. Being an international person brings a perspective that comes from trying to learn everything here and to know you are still learning. He sees the value of integrating voices and the benefits of including equity, diversity, and inclusion in medical care.

Soon after Tolu completed his PhD, his son was born. When the boy was between seven and eight months old, he started running a fever and Tolu brought him home from day care. He took him to his family doctor, who had been a pediatrician for twenty-five years. He examined the baby and said to bring him immediately to the emergency room, as he was sure he had a viral infection. That was where the nightmare began, and Tolu almost lost his son. The ER doctor said it was a bacterial infection and, noting Tolu's accent, he assumed that he had just arrived from Africa and insisted on the wrong treatment. Tolu had to state repeatedly that he had been in Canada for five years, had not caught a disease while travelling, and that his son was truly Canadian. This is when Tolu became aware of racism in health care. He says the problem is that health care is physician-centered and not patient-centered. What if he had not been well educated? What if he had not had the ability and courage to question? Just think . . . The phrase remains incomplete, but Tolu is himself the answer. He is the antidote to stereotypes. He has had amazing mentors, and, in turn, he mentors others. One number at a time, he will contribute to positive change, and he will find the innovative means to effect it through careful and brilliant analysis of data, seeing patterns that others miss and proposing innovative solutions that help everyone.

59

MARIE-ODILE JUNKER

Creating the Tools for Learning

Ottawa, ON

Innovation may mean identifying a challenge that had previously been ignored or adopting new tools and techniques to solve problems. It may mean envisaging an entirely different approach and it may mean redefining the research community.

Marie-Odile Junker's work is innovative on all levels. A linguist, she first observed several decades ago that instruction in Indigenous languages was not widely available across Canada, and she wanted to contribute to improving the situation. She began by learning herself, listening to native speakers and recording their words. She created what she calls a participatory action research framework and collaborated with Indigenous scholars and members of their communities, who directly contribute to the collection of knowledge and understanding its application. All of her work is co-created and co-authored with Indigenous colleagues.

Marie-Odile works with and for Indigenous communities and innovatively uses the internet and computers to compensate for distance, inviting Indigenous language speakers themselves to join the research team that defines both the goal and the process.

In short, Marie-Odile created with members of the Cree and Innu communities a number of collaborative websites for Cree and Innu languages, an

Algonquian Linguistic Atlas, a digital infrastructure for Algonquian language that includes the contributions of eleven participatory dictionary teams, and many communities of speakers who add words themselves. At the end of 2021, the Atlas featured 47 languages and dialects, 67 speakers, and over 25,000 sound files. The Innu dictionary, in its first year alone, had 11,000 words searched and now has over 250,000 each year. All the participating dictionaries together now average a million words searched each year.

This work helps restore to Indigenous peoples the languages they lost when children were removed from their communities. It contributes to their identity and their ownership of place.

The work is carried on in many ways by many brilliant scholars. For example, Dr. Chantelle Richmond, Biigtigong Anishnaabeg, a professor at Western University in London, Ontario, has canoed with her students down the Biigtig Ziibii river, renaming the Pic River in Anishinaabemowin, and reclaiming the land and their heritage. She is taking the next step and returning the language to the land.

But let us look at how the ever-curious Marie-Odile, born in a small town in Alsace, France, ended up canoeing in Canada and working to preserve the languages and cultures of the Indigenous population.

As a child, she was fascinated by everyone and everything she encountered, from people who travelled through Alsace, to the animal and plant life in the meadows and forests around her home. When her mother would ask what she wanted to become one day, she would reply, "A traveller." As she grew, she began to think she might also like to study biology and explore the world to discover new species, and on entering the Université de Strasbourg, she enrolled in biology. Two months later students were expected to dissect frogs and mice, a process she found repugnant. She thought there must be ways, other than torturing small animals, to access scientific information and, walking across campus, happened on the Théâtre de la Liberté, where a play by Goldoni was being presented. She entered and discovered a new passion. She changed her major to Arts (Lettres) and quickly realized that she had a talent for linguistics.

While her father was not very happy at the change, because he had always thought Marie-Odile would be the scientist in the family, he supported her choice, and Marie-Odile continued her studies. In the summer of her second year, she went with a Greek classmate to the island of Crete, where they set up a school in her friend's family garage. They both returned to their studies and her friend later became a successful entrepreneur, opening a private school in Crete. Marie-Odile completed her degree and a certificate in teaching French as a second language at the Sorbonne in Paris, and then returned to the Université de Strasbourg for a master's degree, at the same time teaching French to foreign students and Vietnamese refugees. Since she wanted to learn English, she signed up for a one-year exchange with the University of Houston, Texas, where she taught French to her students and learned English herself. She then taught for a year in a French high school while continuing graduate studies and, at the suggestion of a professor, wrote an article based on her master's thesis. This led to an invitation for a postdoctoral fellowship in Canada (actually, predoctoral, in Marie-Odile's case), which was followed by a trip backpacking in South America, during which she came to the decision that she would return to live in Canada. While she loved travel, she felt truly at home in Canada. She then completed her PhD at the Université de Sherbrooke, and also learned to canoe, discovering a new passion. She was an active volunteer in the environmental movement and met her husband, whose work involved preparations for the Rio Earth Summit, and Bella Abzug, the feminist lawyer, retired US senator, and a leading ecofeminist with the United Nations at the time.

Marie-Odile started learning Ojibway in the Friendship Centre in Ottawa and realized that it had few resources. She decided she could truly serve the population in Canada by contributing her talents to the study and preservation of Indigenous languages. She went to the Centre for Indigenous Research and Culture at Carleton University, where she was given a $600 tape recorder to record and study the Cree and Ojibway languages.

She also became involved with an interdisciplinary group at Carleton University that supported participatory action research. With linguist Marguerite MacKenzie from Memorial University, she visited far-flung villages, creating an online grammar and dictionary that would be augmented by the community members themselves and be an aide for teachers and students alike. The work has been arduous and there were times when Marie-Odile wondered if she had taken the right path. After all, language lies at the heart of culture and is both personal and something that belongs to the community of speakers. It is fraught with politics and difficult emotions, as well as the challenges posed by the small size of the population, the number of languages and dialects, and the expansive and difficult terrain across which people are spread.

Marie-Odile's aha moment occurred in a dream in which she was walking in what seemed to be northern Canadian wilderness. She was feeling sad about the fact that Indigenous languages needed support. It was winter and she felt very cold. She came upon a teepee on a hill and knew there was a fire inside because she could see the smoke furling upward from it. The entry flap opened, and a voice said, "Come in, my child, and warm yourself." She entered and observed people sitting around a fire, dressed in the traditional clothes of several tribes from different periods in history. They asked if she was cold, and she said she was, and added that she was also sad. The voice said, "But you are a linguist. What can you do?" In her dream, Marie-Odile replied that she did not know. An old woman then gave her an empty basket and said, "You will know what to do when the time comes."

A year later, in Waskaganish (formerly Fort-Rupert), Marie-Odile was working on the East Cree grammar and dictionary project with Marguerite MacKenzie and an Indigenous woman named Daisy Moore, who was working to document her language. Unprompted, Daisy told her about a strange dream she had in which she was walking on the land picking blueberries and putting them in her skirt, which she held up in front of her. It was summer, and as she gathered the plump berries, she was thinking about what she had done in her life with her friends, as they had worked so hard to document their language. She looked up, saw a teepee with the flap open, and she stepped in, but as she did, she stumbled and the blueberries tumbled out of her apron, rolling down the hill. A voice said, "My child, you must find a basket for your blueberries."

Whenever Marie-Odile feels discouraged about the immensity of her project, she remembers the matching dreams. She says that when a vision takes hold of you it is both a blessing and a curse. She realized that her dream represented her project, and whether you believe in dreams or not, its metaphorical application to her work was undeniable and the matching dreams, either predetermined or serendipitous, were symbolic of the work that needed to be done and the collaboration required. Many groups of Indigenous Cree women have assiduously gathered words and worked together to create these dictionaries and tools that are accessible and well used by the communities.

Marie-Odile's efforts have been recognized with the Governor General's Innovation Award, and she has shared numerous awards with Indigenous scholars for the important work they have accomplished together. She says that language continues to evolve, and the Indigenous population is expanding. With it comes a growing interest in learning. Marie-Odile looks to the future and says the dream will remain *inachevé* (unfinished) until the precious basket can be passed along to a passionate Indigenous scholar or, preferably, an entire team!

60

FRASER TAYLOR

Reinventing the Atlas

Ottawa, ON

The inscription over the entrance to the small schoolhouse in Leven on Fife read *Wha Tholes*. "Never Give Up. Persevere." The schoolhouse no longer stands, and the inscription now hangs over the lintel in a private home, which replaced it. But for Fraser Taylor, the lesson he read every day as he entered the schoolhouse remains ever fresh in his memory.

Leven on Fife was the home of the Taylors, a family of such very modest means that Fraser would never have been able to attend university had the government not made education free just in time for his matriculation. Before the change in legislation that provided free education after the age of fifteen, Fraser says he would have been lucky to remain in school after the age of fifteen. Growing up speaking English in an area where Lowland Scots was spoken in the streets, however, made his life difficult with his peers.

Fraser owed his English to his mother, who was a self-taught musician and instructed him in speaking and in singing. She entered him in national song and elocution contests. Her encouragement gave him a lifelong appreciation for music, as well as the linguistic skills he needed to succeed at university.

Fraser took a variety of summer jobs. He was foreman of a Gaelic-speaking crew on the River Tay. He was also an agricultural laborer digging potatoes, but by far his most amusing job was chasing poachers. He recalls walking along the high street of the coastal town of Whitehaven and hearing splashing under the bridge, and upon peering down, he saw two naked men, each with a long pole with a forked end, attempting to poach salmon. When they realized they had been caught they threw down their poles and took off running without their clothes down the busy, Saturday morning streets. Fraser did not catch up to them, for laughing at the expressions and reactions of the shoppers. On other occasions he did catch the miscreants and the stories he tells revolve around the sometimes uneven and droll applications of the justice system of the day.

At the University of Edinburgh, he decided to do his PhD on rural development in Africa. This required extensive fieldwork, and there were no research funds available, so Fraser took a job with the British Colonial Office and the Government of Kenya. In preparation, he spent a year at the University College London Institute of Education, studying education for tropical areas, and learned Swahili at the School of Oriental and African Studies. He was posted in Kenya to the only high school in an isolated rural area and became aware of the incredible disparity in the world. Of forty thousand primary school students, only thirty would be selected for high school. The competition revealed to him the priorities of the population and the incredibly high hopes an entire village would place in one scholar. Education was the only means of social mobility.

Fraser also learned the local language, Kikuyu, and was able to understand the motivations and value system of the region. For example, when offered a choice: A, B, or C, the local people would come back and say none of the above and propose a combination of the three so there were no winners or losers, only consensus. After five years in Africa, he returned to Edinburgh to defend his thesis. He based his theoretical findings on data he collected listening to the people of his educational district in Kenya. He also enlisted all his local students to assist in collecting data for him. He likens this to what is now called crowdsourcing.

Fraser then accepted, in 1966 to '67, a position at Carleton University, a new university in Ottawa, where he was determined to be innovative and create a truly interdisciplinary learning experience. His methodology was influenced by his fieldwork in Africa. He would listen first and use a bottom-up approach to program and course design. He taught his students to listen, as well. He wanted to eliminate preconceived ideas, and he involved his classes in projects where they connected with remote communities, bringing people together to tell their stories. Fraser says the technology he used was interesting, but not sufficient. It is *how* you use technology that is most important. His work is about people, not technology.

If you were to seek out Fraser's work on the internet, you would not find a typical atlas. He is the originator of cybercartography, which he introduced in 1997. A "cybercartographic atlas" is the term he devised to describe an atlas that includes all kinds of qualitative and quantitative information linked by location and presented in multisensory, multimodal, and multimedia formats. It tells the stories of local people from their own perspective and includes information rarely, if ever, found in a traditional atlas. The atlas features audio components, with Inuit telling their history, their stories of the land in their language, with English and French translations. Fraser says for the Inuit, "every journey is a story, and every story is different. The name of each place is important, but even more important is the story of how the place got its name." These are, along with the Atlas of Ottawa Hull, among the first such atlases ever created, and Fraser is widely regarded as one of the pioneers in the application of computers to cartography.

Fraser, with his postdoctoral student, is now collecting the stories of residential school survivors to make them accessible to students across the country so they can be heard and understood. He has broadened his scope at the same time and is working with communities in Brazil, Mexico, and South Africa. Each region has an individual project, and each is innovative. All projects are related to local initiatives controlled by the local population.

Fraser initiated the Barbara Petchenik Children's Map Competition while he was president of the International Cartographic Association. For each of the last thirty years there has been an international competition to collect and share maps drawn by children of how they see the world. These maps and the stories they tell can be shared around the world, and they are all saved in a public archive in the Carleton University library.

Current projects in Mexico include mapping local observations of problems including health, the environment, and socioeconomic issues. Stories of the sky and legends of volcanic eruptions offer clues to the future as well as the past. In Brazil the population in the northeast, a landscape of sand dunes and lakes, is creating a map that shows the environmental and economic opportunities and challenges of the region.

With each project, new proposals arise from the local population. When Fraser and his team completed the Cybercartographic Atlas of Arctic Bay in Nunavut, all twenty-six communities in the region wanted to create their own atlas. The Arctic Bay Atlas includes a rap video titled "Don't Call Me Eskimo," a description of the many social and economic challenges facing local youth. The video went viral on YouTube. Ever innovative, Fraser worked with Arctic College to develop a course that would encourage residents to become active research creators on their own.

Understandably, Fraser cannot identify a single eureka moment in terms of his involvement with local communities. He says one thing constantly leads to another. There is, however, such a moment with cybercartography when he introduced the concept to over one thousand of the world's leading cartographers in his keynote address to the International Cartographic Conference in Stockholm. He credits his time in Africa for opening his eyes, his mind, and his heart. He says many of the theories of international development he had read previously were wrong. The information he found, his data points, did not fit the theories. At first he thought his data must be wrong, but it was the theories that were wrong. Theory must be conceived from another perspective that comes from close contact, listening, and understanding others. Good research requires curiosity, innovation, and ideas. Too much specialization does not lead to innovation. One needs a holistic approach with international experience, and one must learn the language of the people to do good fieldwork and listen and learn from the people, and they, in turn, will play an important role in developing projects.

The best advice offered by this award-winning scholar, whose work embraces the world and brings everyone together in understanding, comes from the small seaside town where he was born. *Wha tholes*—persevere.

Conclusion

There is no end to innovation. There is always a new beginning. Hardly a day goes by that we are not faced with additional challenges that were not there yesterday. When there are none in sight, we can always think of a better way of doing whatever we are already doing.

As soon as researchers created a vaccine against tuberculosis, reporters were asking about malaria. When an engineer invents a way to perform medical procedures at a distance, we will need to find a way to manufacture and distribute that equipment around the world. When we have listened to Jeremy Dutcher's recording of Indigenous music, we will be ordering the next recording before he has even had the time to write the music. When we solve the crises in immigration, housing, and carbon emissions, a new list of issues will doubtless await our attention. An innovator's work is never done. We will always need good, new ideas and the people with the knowledge, skills, and passion to see them into existence.

It has been a privilege to write about some of the extraordinary people who live and work or who have lived and worked in Canada. I found them truly inspiring. They gave me reason to hope for the future and to believe not only that we can do better, but that we actually will.

I thank all the people you will meet in the pages of this book—and all those who deserve inclusion and who await a future collection.

I am grateful to you, the readers, who are hopefully not reading this final page first, but who will not have been able to stop enjoying the stories of these fascinating people who are contributing to a better world right from the beginning.

And I dedicate the book to future generations of innovators in every field who will perhaps be inspired by these trailblazers, who defied both the odds and those who tried to discourage them. Their curiosity and passion withstood the test of time, and now your challenge will be to build on the accomplishments of all those who have been innovative and who are still innovating today.

Acknowledgements

Thank you to the Board of the Foundation for Innovation for encouraging me to pursue this project in my spare time and to my friends and colleagues Denise Amyot, Mike Tremblay, Jo-Anne Poirier, Bernard Leduc, Jacques Frémont, Chantal Beauvais, Gregory Pilsworth, Elizabeth Shilts, Jodi Di Menna, Suzanne Nothnagle, Michael O'Reilly, Martha Crago, Marie-Josée Hébert, Meg Beckel, Anoush Terjanian, Azra Yousef, Danielle Fremes, Michael Strong, Iain Stewart, Ted Hewitt, Harsha and Sudha Dehejia, Leslie Weir, and Linda Savoie.

Special thanks to Jacques and Donna Shore, Bryan Jones, Pat Bailey, Ian Rankin, Pierre Normand, Rafik Goubran, Anna O'Reilly, Louise Paul, Charmaine Dean, Alice Aiken, Bill Ghali, Karen Mossman, Dugan O'Neil, Kevin Smith, John Matlock, Julie Cafley, Lise Bourgeois, Lyse Marchand, George Addy, Suzanne Gumpert, Jean-Marc Carisse, Lesley Rigg, Nancy Ross, Manav Ratti, Tobias Calderini, and Nicole Rochford.

Very special thanks to everyone included in this volume for their participation and suggestions, and to all your assistants who made it possible for us to get together!

The biggest bouquets go to Kevin Hanson for reaching out to me, for the ideas and encouragement, and to Janie Yoon and Jim Gifford, not only for being great editors but also for being kind, thoughtful, and creative.

The Canada Foundation for Innovation funded many of the research projects noted in this volume, and I am grateful to the researchers, the foundation, and the government of Canada, the governments of all the provinces and territories, industry, and the institutions for supporting the work that has led to such brilliant innovations.

Index